The School Projects Idea Book

by Michele Davis

Los Angeles, CA
www.pgpress.com

The School Projects Idea Book

ISBN: 1-931199-29-9

This book, and all titles in the Parent's Guide series, are available for purposes of fund raising and educational sales to charity drives, fund raisers, parent or teacher organizations, schools, government agencies and corporations at a discount for purchases of more than 10 copies. Persons or organizations wishing to inquire should call Mars Publishing at 1-800-549-6646 or write to us at ***sales@marspub.com***.

At the time of publication of this book, all of the information contained within was correct to the best of our knowledge.

Please contact us at *parentsguides@marspub.com*.

Printed in Canada.

Edwin E. Steussy, CEO and Publisher
Anna-Lisa Fay, Project Editor
Dianne Tangel-Cate, Editor
Lars H. Peterson, Acquisitions Editor
Michael P. Duggan, Graphic Artist

PO Box 461730, Los Angeles CA 90046

Contents

Chapter One

Introduction

The School Project Ideas Book is conceived to help public, private and homeschooling teachers use real-world projects as part of their standard teaching practice. The projects in this book are designed to develop students' critical thinking and problem solving skills, as well as teaching them an important subject, be it math, science, language, social studies or something else entirely.

By taking a part of the Real World and having students work out answers to real problems, they gain insight into the universe around them. Each sample project developed in this book can be modified to suit virtually any classroom environment. And each project can be expanded or contracted, depending on the individual abilities and interests of the students.

The possibilities are endless, so let's get started.

Chapter One

How to use this book

This book is divided into three sections. The first section explains the structure, assessment and evaluation of projects. The second gives tools and tricks to obtain the greatest educational gain from each project and the final section consists of thirty fully developed projects based on guidelines outlined from the Buck Institute of Education in Novato, California. The institute researched and established criteria for Project Based Learning (PBL) and a handbook can be obtained from their website at www.bie.org/pbl.

Ready, Set, Go

What you do at conception will determine the success of execution, so please take extra time at the beginning to set the tone, build the foundation and make sure everyone is clear on the project, its outline and outcome. Guidance is needed to establish expectations and to set students in the right direction and once this groundwork has been established, the exercise is wide open. The greatest flexibility comes from within the structure that you create out of your chosen project and the activities which subsequently follow vary, depending upon the needs of your students and the curriculum. It's important to differentiate each project slightly to meet specific needs.

Structure

All projects have the same structure and are divided into three distinct stages, beginning, developing and concluding. Follow the same sequence of steps for every project, creating continuity and repetition so students can begin to tap into the thought process behind the ideas. The process consists of the following steps.

Vocabulary and Learning Outcomes

Incorporate any vocabulary specific to your study and adjust the learning outcomes. Add or take away as needed.

Brainstorm and Discuss Subject

This is a free-flow of communication which establishes the basic knowledge students need to proceed. Now is a good time to ask questions and invite students to guess or share any relevant information. Record the big ideas, concepts and vocabulary. You can use graphic organizers to represent connections in information and write on single transparencies or chart paper to make the information easily accessible. These 'chat' sessions allows you to assess the thinking of your students and quickly evaluate which students have baseline knowledge and which students have no prior experience.

Identify Project Focus

As the brainstorming session ends, guide the discussion toward the project focus which lies at the core of the subject. All explorations lead to and from it. This is the main purpose underpinning the root of the matter. Guide your students to encapsulate it into one clear sentence or goal, or introduce the focus and display it clearly. Ask students to record it in their journal for future reference.

Identify Learning Objectives and Explain Assessment

After detailing the focus, explain how the project will be assessed. Decide on the rubric (evaluation method) needed for assessment at the end of this chapter and present it to the students. Identify specific key objectives and give students the opportunity to ask questions. This refines their thinking on how they will be assessed.

Create Project Journal

The first activity is to create a 'master' project journal for each student which is used at every stage. All information about the subject is recorded in the journal and includes initial chat session, facts, websites, books, encyclopedia data, drawings, discussions, interviews, ideas and questions. The journal can be a store bought folder or made from construction paper and paper stapled together and includes both blank paper and graph paper for drawings, graphs, and tables. Creating a journal with pockets is ideal for adding handouts and creating a place for students to put their journal each day. Don't allow the binder to be taken out of the classroom, since this creates the potential for a lost journal.

Record Project Focus, Chat Session and Estimate End Product

Students record the project focus at the top of page one in their journal. At the bottom of the page, students estimate their end product choice. At times, students have a choice in the final product, other times you will dictate their choice and decide based on the needs of your students and the curriculum goals. Establishing each of these in the journal creates a direction to follow and the space in between these two is used for questions and planning. On the second page, students record the brainstorm or chat session. Ask them to copy everything recorded during this session and this makes the basic knowledge readily available to them.

Create Project Plan

Directly under the project focus, students write five questions about the subject. Each question serves as a starting point, which will in-turn require research. Student's record where they will first look to begin their exploration and you can hold a brief conference with each student or group before they conduct fieldwork and ensure their thinking is on track. Also verify their research avenues as valid and reliable as some students are quickly able to create a plan and get started, where other students need guidance each step of the way.

Chapter One

Conduct Fieldwork

Students conduct fieldwork by talking to professionals, performing research on the internet, watching TV, reading books, magazines, and newspapers. Each episode is cited in their journal and you can conduct mini-lessons on citing information on the first research day and on subsequent days as needed. Show students how to make lists, draw relationship webs or graphic organizers and guide them to choose key elements of information to help them plan out their research.

Students must delve into the subject deep enough to complete the activities but monitor themselves to not get off track. The project focus, plan and journal are invaluable tools to keep them focused.

In the beginning, ask every 10-15 minutes for students to check their direction. This helps build a habit of staying on track and not going off on a tangent. By referring to the focus, plan, and end product, students can verify if they have deviated from the focus and get back on track quickly.

Occasionally students come across information that is not relevant to the project but is information they have been looking for otherwise. Depending on the age and responsibility level of the students, ask them to take a page at the back of their journal to record brief information of off-topic subjects. Allow only one page for this.

Conduct Activities to Learn

At each stage, activities cater to varied ability levels and learning styles. Choose or change the activities to differentiate instruction and learning and include all or most activities to create the greatest depth. The beginning activities build the foundation of knowledge. The developing activities personalize learning or relate students more closely to the subject and the concluding activities are end product choices. While conducting fieldwork, students may come across activities or ideas to try. Refer to the project focus and end product choice to judge the applicability and feasibility of their suggestion.

Create End Product Demonstration

Each of the concluding activities is an end product choice. Students should not restate information but utilize their new knowledge to create a demonstration of what they learned.

Possible end choice includes:-

- Poster displays
- Building a website
- Dioramas
- Written reports
- Graphs/surveys
- Product prototypes – made with cardboard, metal or plastic scraps, Plaster of Paris etc.
- Board, paper, or card games
- Picture/photo displays
- Slide show presentations
- Timelines
- Graphic designs – created by hand or computer
- Maps – drawn by hand or computer

Results will vary. Sometimes students delve deeply into the subject, make great generalizations, form fantastic final products and develop critical thinking in the process. Other times, the end product doesn't turn out as expected, has different results or just doesn't work. This is the essence of learning; scientists, CEO's, construction foremen and a host of other professionals try things that don't work. The process is not about being 'right,' it is about the journey they embarked upon from A to B and what was experienced between these two points. Regardless of right or wrong, the end product should meet all predetermined assessment criteria.

Reflection on Project

Captured through journal entries, a final chat session asks students to summarize their learning. Students write a paragraph or list of what they learned and discuss what surprised or challenged them and what was of particular interest. Each written reflection is shared during our final chat session and students can be asked to share their entries out loud to the rest of the class.

Chapter One

Planning

Projects offer teachers an authentic teaching practice that caters to all students' needs. However, using projects requires a certain amount of pre-planning and the more projects you complete, the easier the planning process becomes. Projects also offer educational rewards far beyond worksheets and book questions and you can use the tools and tricks included in this book to find the greatest educational gain from all the projects.

Evaluation

Before embarking on a project, establish an effective plan allowing you to create effective assessment tools. The questions below will help establish guidelines and answer these questions at the outset of each project.

- Do the questions take the direction you want to take?
- Which direction would better enable you to meet the needs of your curriculum?
- Are you deciding the direction or are the students, or a combination of both?
- Could other disciplines be included with the project?
- Are there specific skills or objectives you would like to add to the project?
- How much time do you have to complete the project?
- What resources are available for students to use?
- Are there any specific sensitive areas for this project?
- Are there any government entities that have information on the topic?

Assessment

After determining the direction of the project using the questions above, chose the objectives and skills unique to that direction. Rubrics are the most amiable assessment tool and since students may study in different directions at the same time, a flexible tool is necessary. A rubric is an outline or grid of criteria and defines what specific objectives and criteria will be assessed. It specifies the expected learning outcomes and other suitable factors such as neatness, creativity, and appearance. Use the rubric at the beginning stage to introduce expectations and guidelines, and you can also refer to the rubric during the exploration to maintain focus, and after completion to score the overall project.

There are two basic types of rubrics: holistic and analytical. A holistic rubric assesses an overall project or activity and gives one total score. An analytical rubric handles specific skills; each receiving its own score. Each individual score can stand alone or become part of an overall project score.

Holistic Rubrics

Holistic rubrics are good for assignments that receive one grade because there is one main objective. Below is an example of a holistic rubric for a timeline, obtained from 'The Staff Room for Ontario's K-12 Teachers' website at http://www.odyssey.on.ca/~elaine.coxon/rubrics.htm.

	Excellent	Good	Fair	Poor
Content	People, events and inventions on timeline are important. **and** Timeline is not cluttered. **and** Spelling and grammar are correct.	People, events and inventions on timeline are important. **and** Timeline is somewhat cluttered. **or** Spelling and grammar are incorrect.	Incomplete information on people, events or inventions on timeline. **or** Timeline is somewhat cluttered. **and** Spelling and grammar are incorrect.	People, events and inventions on timeline are not important or are missing. **and** Timeline is cluttered. **and** Spelling and grammar are incorrect.

Analytical Rubrics

Used to assess specific parts, portions, or assignments within a project, each portion is assigned a score. The scores can then be totaled to determine a complete score for the entire project. This research rubric was obtained from the Teach-nology website at www.teach-nology.com.

Chapter One

Generic Elementary Research Rubric

Name: ______________________________

Teacher: Mrs. Davis

Criteria

Points

4

3

2

1

Introduction/ Topic

Student(s) properly generate questions and/or problems around a topic.

Student(s) generate questions and/or problems.

Student(s) require prompts to generate questions and/or problems.

Questions or problems are teacher generated.

Conclusions Reached

Numerous detailed conclusions are reached from the evidence offered.

Several detailed conclusions are reached from the evidence offered.

Some detailed conclusions are reached from the evidence offered.

A conclusion is made from the evidence offered.

Information Gathering

Information is gathered from multiple electronic and non-electronic sources and cited properly.

Information is gathered from multiple electronic and non-electronic sources.

Information is gathered from limited electronic and non-electronic sources.

Information is gathered from non-electronic or electronic sources only.

Summary Paragraph

Well organized, demonstrates logical sequencing and sentence structure.

Well organized, but demonstrates illogical sequencing or sentence structure.

Well organized, but demonstrates illogical sequencing and sentence structure.

Weakly organized.

Punctuation, Capitalization, & Spelling

Punctuation and capitalization are correct.

There is one error in punctuation and/or capitalization.

There are two or three errors in punctuation and/or capitalization.

There are four or more errors in punctuation and/or capitalization.

Total——>

Teacher Comments

Chapter One

Creating Rubrics

Writing your own rubrics allows you to address objectives specific to your exploration. Use the questions below to create a clear understanding of the objectives and skills to assess, help you decide your goals, how you will reach those goals, and what you expect of your students.

What is the motivation for the project?
What curriculum goals do I want to meet?
What objectives are natural to the project?
What skills are natural to the project?
Can any other disciplines be included?
What overriding ideas should students internalize, if any?
Is there a "take home message," such as an environmental or humanitarian message?
What specific skills will be taught?
What activities will develop these skills?
How are the students expected to demonstrate their learning?
How will the learning be measured?
Do any students have specific needs that should be addressed prior to beginning?

Rubrics on the Internet

http://rubistar.4teachers.org/

This website allows you to write the criteria and generates the full grid for you.

http://4teachers.org/projectbased/checklist.shtml

This site allows you to create rubrics for your specific needs.

http://www.teach-nology.com

This website provides great background for developing rubrics as well as a rubric maker.

Tools and Tricks

Projects can be used to decrease time needed to teach and expand knowledge of many subjects. Using reliable methods and tools is essential to achieving all the project goals within a definite timeframe.

Using Effective Questions

To get the most out of projects, immerse your students in information about the subject. Read and explain obscure facts and build graphs out of all sorts of information. By exposing students to the topic, it helps to inspire curiosity and thinking. The goal is to enthuse students about the exploration, opening a door to expose our complex world. Understanding these complexities requires critical thinking and problem solving abilities and asking questions develops these skills. At the developing and concluding stage, the questions and activities require higher cognitive thought.

The following adaptation of 'Bloom's Taxonomy' by John Maynard is an invaluable tool to develop higher thinking. The chart is divided into sections and displays examples of question beginnings at each cognitive level. While moving through the six levels, the depth of cognitive skills expands. Each question beginning for the first four levels requires deeper thought. Use the question beginnings as a guide when constructing your own questions. Familiarize yourself with the chart and vary your questions to elicit the highest cognitive thought for your students and differentiate question beginnings to address specific ability levels and needs.

Chapter One

Bloom's Taxonomy

Model Questions and Key Words

Based on Bloom's Taxonomy, developed and expanded by John Maynard

I. Knowledge (drawing out factual answers, testing recall and recognition)

who	where	describe	which one
what	how	define	what is the best one
why	match	choose	how much
when	select	omit	what does it mean

II. Comprehension (translating, interpreting and extrapolating)

state in your own words	classify	which are facts
what does this mean	judge	is this the same as
give an example	infer	select the best definition
condense this paragraph	show	what would happen if
state in one word	indicate	explain what is happening
what part doesn't fit	tell	explain what is meant
what expectations are there	translate	read the graph, table
what are they saying	select	this represents
what seems to be	match	is it valid that
what seems likely	explain	show in a graph, table
which statements support	represent	demonstrate
what restrictions would you add		

III. Application (to situations that are new, unfamiliar or have a new slant for students)

predict what would happen if	explain
judge the effects	select
what would result	tell what would happen
tell how, when, where, why	tell how much change there would be

IV. Analysis (breaking down into parts, forms)

distinguish
identify
what assumptions
what motive is there
what conclusions
make a distinction
what is the premise
what ideas apply
what's the relationship between
what's the main idea, theme
what literary form is used
implicit in the statement is
what is the function of
what's fact, opinion
what statement is relevant
related to, extraneous to, not applicable
what does author believe, assume
state the point of view of
state the point of view of
what ideas justify conclusion
the least essential statements are
what inconsistencies, fallacies
what persuasive technique

V. Synthesis (combining elements into a pattern not clearly there before)

create
tell
make
do
choose
develop
how would you test
propose an alternative
solve the following
plan
design
make up
compose
formulate
how else would you
state a rule

VI. Evaluation (according to some set of criteria, and state why)

appraise
judge
criticize
defend
what fallacies, consistencies, inconsistencies appear
which is more important, moral, better, logical, valid, appropriate
find the errors
compare

Chapter One

Using the Jigsaw Method

The Jigsaw Method is a learning activity that builds community and cooperation in your classroom and enables you to increase exposure to content but decrease time needed for students to synthesize it. This method requires students become experts in a specific area of study and then share their learning with other students. There are many benefits to using this approach. One is the need to cover many directions in an exploration at one time. You can assign different activities to different groups and this creates an appreciation of the big picture for all students. Using the Jigsaw Method also allows you to choose activities for various abilities and learning styles. Tailor activities for each student or group based on their specific needs. More information about implementing and using the Jigsaw Method may be obtained at www.jigsaw.org.

Using Crash Courses

Crash courses may be used for the whole class, small groups, or individual students. Use graphic organizers, charts, lists, and note-taking to help students internalize the details. Use crash courses to re-teach a skill, demonstrate particular concepts, or provide help in documenting facts.

Jump Right In

Working with projects provides opportunities to learn in authentic formats. Read a project and make it your own. Change any aspect you see fit. Each project is limited only by your imagination. By learning this teaching approach and using the projects in the book, you will save time, meet more curriculum objectives, and address the varied needs within your classroom. Remember, experience is the best teacher so jump right in and enjoy the journey!

Chapter Two

Social Studies Projects

Grades 2-3

Flags Unplugged

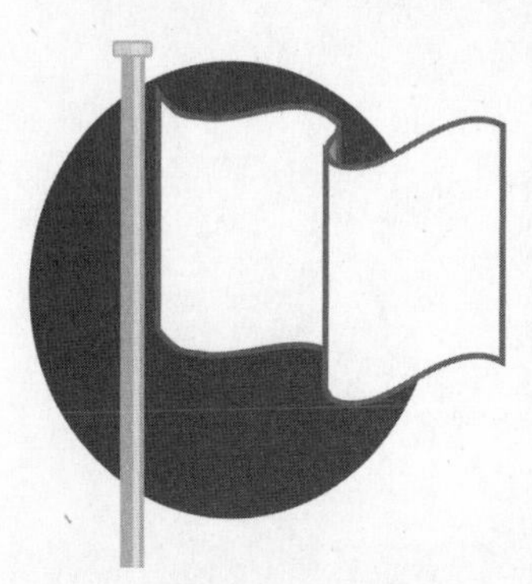

Flags are universal symbols seen throughout the world. All students have been exposed to flags in some form. Understanding flags and their symbolism help the children understand their world and the symbolic meaning behind emblems. By designating symbols and emblems that have a particular significance, students learn about themselves.

In the beginning stage, students work to develop an understanding for the many types of flags. They will also soak-up the history and development of flags in particular states or countries. Each opportunity affords students the chance to study flags specific to a certain subject or area of interest. The direction of the project can vary widely at this stage. It will depend upon your particular curriculum needs.

In the developing stage, the class personalizes their learning by deciding emblems and symbols that have significance for them. These may be personal favorites—historical or religious symbols special to them. Students will also note the many symbols and emblems used by corporations, governments, businesses, and a host of private organizations. A great extension would be to challenge students to note the most flags seen in one day.

In the concluding stage, each child creates a personal flag that contains colors, symbols, and emblems that have special significance to them. They'll feel a sense of belonging and connection through this project. Students learn about each other from the flags as well.

Chapter Two

Project Focus

Flags often represent important ideas, concepts, historical facts, etc. Design a flag for your classroom, home or your own personal flag. Be sure to use and note significance of objects and colors to be used in your flag.

Why this project is valuable

Understanding significant historical facts and demonstrating significance of self enables students to relate to the world around them.

Students will learn

- How to solve a problem
- Developing a useable plan
- Flags of the world
- Symbolism of flags
- Significance of flags in history
- Development of flags
- Personal identity
- Personal symbolism

Vocabulary

- Flags
- Symbols
- Symbolism
- History
- Culture
- Identity

Teacher's Tip

This could be a great beginning year activity that would allow you to jump start your social studies curriculum and get to know the students better. Display the flags in the room for all to see.

What kids should know before beginning this project

Students should have some knowledge of

Flags as part of our culture and the history of nations.

Beginning the Project

Brainstorm and discuss

- Types of flags.
- Purpose of flags.
- Symbolism of flags.
- Overall meaning of flags.

Questions to consider

- Are you choosing a particular state or country to study?
- What significant aspects of flags will you study?
- What is your intended outcome?
- Are you specifying if the flag is for school, home, or personal?
- May students choose?

Activities

- Create project journal.
- Research and report on the history of the U.S. Flag.
- Choose a particular flag. Explain the history and significance of each color and object on the flag.
- Draw a succession of flags from one particular state, showing the changes over time.
- Choose a favorite flag and explain why it is your favorite.
- Read aloud the story of Neil Armstrong placing an American flag on the moon. Discuss significance of this act. *Man on the Moon* (Picture Puffins) by Anastasia Suen; Puffin (March 2002)
- Write questions to guide your thinking about what to include on your flag. What questions would you ask yourself?
- Compare questions with a friend.

Developing the Project

Questions to consider

- What are important parts of your life?
- What objects would you use to represent these important parts?
- What are your favorite colors?
- What are important dates in your life?

Chapter Two

Activities

- Go on a symbol scavenger hunt. Locate common symbols. Draw them and tell what they mean.
- Using a blank United States map, draw a picture of each state's flag in the state.
- Compare two flags that use the same colors, note the significance of each color.
- Make a list of important events and people of your life.
- Create a symbol for each of these people and events.
- Make a symbol and color key for your flag.
- Write and conduct a survey of parents asking what flags mean to them.
- Chart findings.

Concluding the Project

Questions to consider

- What materials are available?
- What colors and objects are students using?
- What meaning is behind the colors and objects?

Activities

- Create your own personal flag. Also create a key for colors and objects.
- Write a letter to a friend including a picture of your flag. Tell your friend how you came to develop your own flag and what each part means.
- Use multimedia program to create your flag electronically, include a key.
- Create a chart comparing five different flags from around the world, note significance of symbols and colors.
- Provide a chart of common symbols seen in everyday life, include where they are seen and what each of them means. Provide suggestions for new symbols to use.

Resources

Books

Each of the books below will most likely need to be checked out through a public library.

DK Handbook: Flags; DK Pub Merchandise (June 1991)

Flags of the Fifty States and Their Incredible Histories by Randy Howe; The Lyons Press (November 2002). This brand new book would not be an easy read for this age group but could be read aloud by the teacher and used as a reference.

States Names, Seals, Flags, and Symbols: A Historical Guide by Benjamin and Barbara Shearer; Greenwood Publishing Group (October 2001)

Internet Sites

http://www.flagworks.com/testimonials.html—This site sells flags from all over the world. The pictures of the flags can be used to complete various activities. Many fascinating facts can be learned about the meaning behind each of the flags.

http://www.law.ou.edu/hist/flags/—More information on the history of the U.S. Flags, including pictures from hundreds of years ago.

http://www.crwflags.com/fotw/flags/—Follow the links on this site to print blank black and white outlines of flags from all over the world for students to color.

Flags Unplugged

Michele's Personal Flag

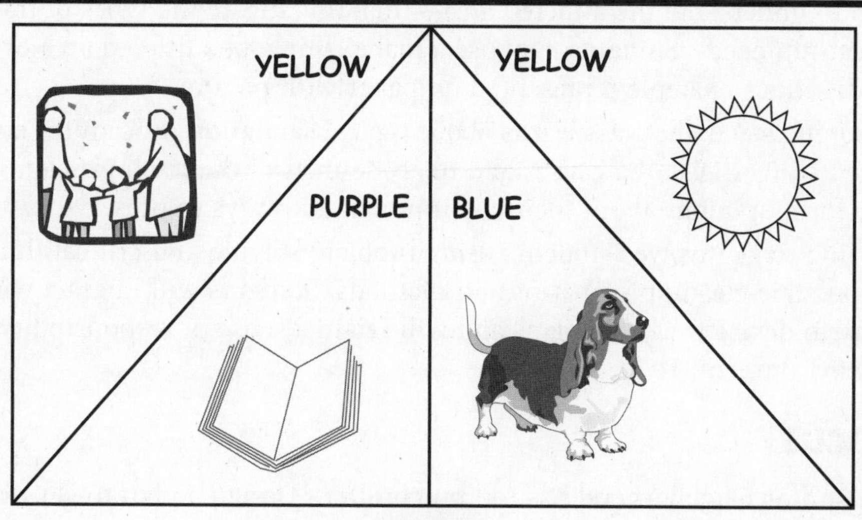

KEY

The Symbols represent the things that I love!

DOG	=	My dog's name is Sweetie
BOOK	=	I love reading.
FAMILY	=	I love my family.
SKY/SUN	=	I love to play outside.

The colors I used are my favorite!

BLUE	=	Sky
PURPLE	=	Royalty
YELLOW	=	Sunny

A sample flag and key.

Chapter Two

Social Studies
Grades 2-3

Too Much Litter

Environmental studies are an exciting topic for most students. Children usually love to be outdoors and interact with nature. They see the beauty of nature and are happy to preserve it. By investigating environmental issues in areas close to home, students learn to notice their environment and initiate a personal responsibility to maintain it. They also learn to monitor their own behavior and develop plans to solve a problem.

During the beginning stage of this project, students will begin to understand the afflicted area—then investigate the types of trash and possible clean-up needs. Students may use a real example of a littered area or they may use a fictitious example represented on paper with pictures.

In the developing stage, the class learns about typical sanitation procedures and processes. Students investigate the causes and likely sources of the trash problems. They'll begin to form opinions about their environment and ways to preserve it.

The concluding stage involves students using problem solving and critical thinking to propose possible clean-up scenarios and methods. Students will interact with their community to devise workable plans and will retain a sense of responsibility to maintain their environment.

Project Focus

Your schoolyard or neighborhood has too much litter. Design a plan to get rid of the litter and prevent it from accumulating again.

Why this project is valuable

All students are consumers. As consumers, they are also trash makers. It is important for children to realize the impact trash and litter have on our environment, as well as how each child, as a consumer, can make a difference in the amount of litter in their school or home community.

Students will learn

- Responsibility for their actions
- How to solve a problem
- Developing a useable plan
- Effects of trash on community
- Different types of litter
- Difference between biodegradable and non-biodegradable items
- How to conduct a survey
- Creating graphs
- Drawing maps

Vocabulary

- Trash
- Litter
- Junk
- Community

Note: These are big words for little kids, use pictures and examples.

- Biodegradable
- Non-biodegradable
- Decompose
- Hazardous—use the skull sign on cans as an example.
- Non-hazardous

What kids should know before beginning this project

Students should have some knowledge of

- What a community is.
- Reading a basic map.
- Looking at top views of maps.
- Types of trash—biodegradable vs. non-biodegradable, large vs. small, hazardous vs. non-hazardous, paper, plastic, aluminum.

Chapter Two

Beginning the Project

Brainstorm and discuss

- Communities—draw a web or graphic organizer to demonstrate properties.
- Litter—different types:
  - paper, plastic, aluminum—recyclable
  - biodegradable vs. non-biodegradable
  - large vs. small
  - hazardous vs. non-hazardous

Questions to consider

- What kind of litter is at the site? Analyze types of trash.
- What is the source of litter—who's putting it there?
- How big is the area that needs to be cleaned?
- Are there any special needs to clean the area—such as machines to carry large items?
- What trash receptacles are currently available?
- What is the normal trash pick-up day?

Activities

- Create project journal.
- Read aloud book(s) about communities, litter, and recycling.
- Journal thoughts about topic—student's opinion.
- Create web or graphic organizer of information, include in journal.
- Draw replica of area including trash placement, include in journal. Using rulers, crayons, markers, pencils and construction paper, students can draw basic outline of area, making sure to include all trash at site.

Developing the Project

Questions to consider

- What community organizations are available to help?
- How could families be involved and help?
- How could the school help?
- Does their project require any special resources or money?
- What would have to happen?
- How would they do it?
- Is there a job for everyone participating in the project?
- What are some possible problems?
- What are possible solutions to the problems?

Activities

- Contact local government for trash pick-up dates.
- Categorize litter and create bar graph on paper or computer showing different types of litter.
- Conduct interviews with community or school members to discuss possible reasons for litter.
- Create an outline of questions to ask before the interview such as:
  - Why do you think there are so many soda cans in the park?
  - Are baseball games played at this park every weekend?
  - Who are the regular patrons to this park?
- With an adult, visit area during a busy time. Look for reasons of excessive litter. Do trash bins fill up quickly? Do people throw trash on the ground regardless of trash bins?
- Create a survey asking other students' possible uses for soda cans, cereal and shoe boxes, and plastic containers.
- Identify other uses for trash.
- Research recycling through books, the internet, or check with local government concerning current recycling projects in the community. Check with art teacher—could some items be reused in art class?

Concluding the Project

Questions to consider

- What is needed to clean the trash site?
- How long it will take?
- How many people are needed?
- What can be done to keep it clean?

Activities

- Write a proposal to the city council or school board suggesting needed changes—it is perfectly OK for it to be on a second or third grade writing level. The point can still be made.
- Develop map showing new places to collect litter, suggesting placement of trash bins.
- Create advertisements for community or school members representing places to put trash.
- Create advertisements for reasons not to litter.
- Make a documentary video on risks of litter in community.
- Create a multimedia presentation on necessary clean-up procedures.

Chapter Two

Resources

- School board or local city government

Books

Recycle That! by Fay Robinson; Childrens Press; 1995

The Great Trash Bash by Loreen Leedy; Holiday House, Inc.; 1991

Recycle! A Handbook for Kids by Gail Gibbons; Little, Brown and Co.; 1992

Internet Sites

www.epa.gov/recyclecity/—This government site is written for kids and has a host of fun activities and information for students to use. Some early readers may need assistance in navigating the site.

www.epa.gov/kids/garbage.htm—Another great site to develop understanding and play some fun games and activities related to the information.

Too Much Litter

Trash Can Proposal

★ = where new trash bins should be.

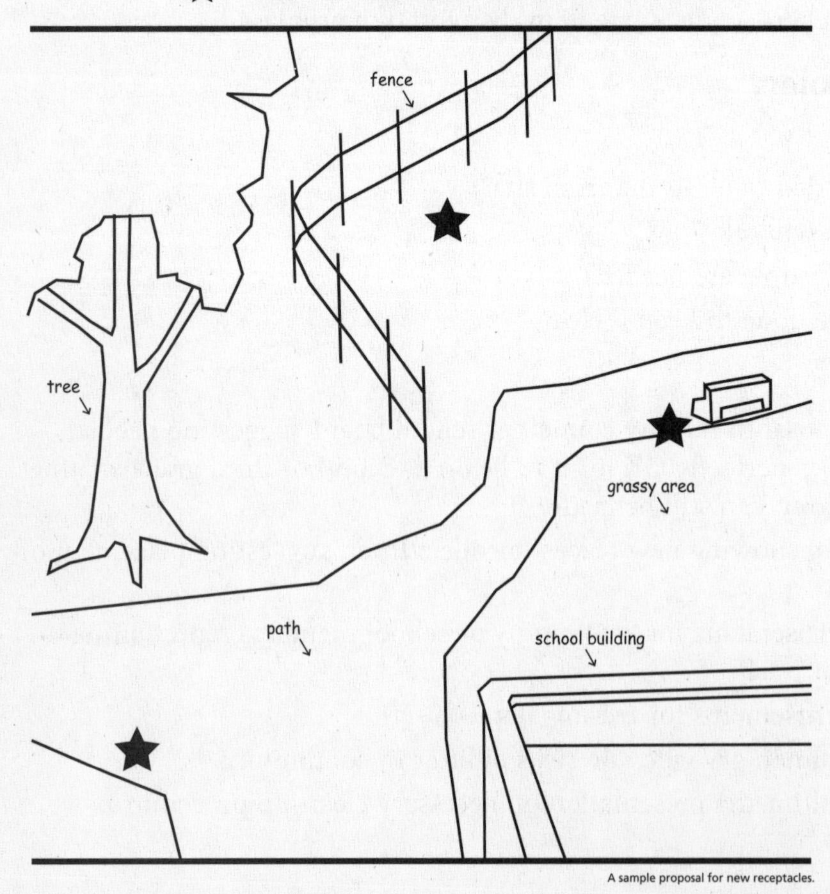

A sample proposal for new receptacles.

Social Studies
Grades 3-5

Family Tree

Families are close to every child's heart. Discovering different family's heritages and customs can open doors to fascinating historical facts. Students are generally excited to learn the facts behind their parents relationship and other family history. Most parents are willing to participate in a project of this sort, but keep in mind there is the possibility of uncomfortable issues that might arise during this discussion.

In the beginning stage, students gather background information about their family. They'll learn the basic structure of families and the differences prevalent in modern families. This is a great opportunity for students to hear old family stories or learn the history behind common traditions.

While developing, each child looks at an overview of the family for as far back as they are able to trace. Students will also learn about the census conducted in the United States every ten years. By studying census data, they learn how facts are obtained about families and what a vast resource this data can be.

During the concluding stage, students will gain a sense of belonging from understanding their family tree. Even children in unique situations will feel a part of a larger community and family. They'll learn about heritage and the many facets of history tied to their family.

Project Focus

You've decided to become an ancestor detective. Create a working model of your family tree for as far back as you can go. Include relevant historical notes when possible. For adopted or foster children, study your current family's history.

Chapter Two

Why this project is valuable

Students will learn

- How to solve a problem
- Developing a useable plan
- Genealogy
- Evolution
- Historical perspective
- Changes over time

Vocabulary

- History
- Family
- Genealogy
- Genealogist
- Surname
- Birth
- Death
- Census

Teacher's Tip

I suggest sending home a note explaining this upcoming study. Express in the letter your wishes to honor all families and their heritage. Allow at least a week for parents to contact you with special concerns. Work with parents to address any and all special concerns.

What kids should know before beginning this project

Students should have some knowledge of

- Student's birth date.
- Where they were born.
- Parents full names.
- Family's heritage.

Beginning the Project

Brainstorm and discuss

- Dates of students' births.
- Favorite family stories.
- Heritage of families.
- Various ethnicities.

Questions to consider

- Do you have students of varied ethnicities in your classroom?
- If so, what makes them different/the same?
- Are any students foster children or adopted?
- Any special considerations for this project?
- Are pets considered members of the family?

Activities

- Create project journal.
- Record full names of all individuals in your family.
- Read aloud books about families. Discuss differences/similarities.
- Draw a t-chart showing similarities.
- Pick a favorite cartoon or TV show. Draw a relationship map of characters involved.
- Ask parents to tell students a favorite family story.
- Write parent's story in your own words.
- Illustrate story and publish a copy for parents.
- Write a survey asking family members favorite food, color, music, etc.
- Conduct survey. Categorize family members according to favorites.

Developing the Project

Questions to consider

- Have students discovered anything interesting about their family?
- Are parents cooperative?
- Do students feel comfortable with subject?
- Is information retrievable and useable?

Activities

- Fill in Family Group Record Page from www.rootsweb.com website.
- Explain if this was easy or difficult and why in your journal.
- Pick a person from your family tree and investigate the history behind their name.
- Conduct research on the Census in the United States.
- Explain what a census is in your journal.
- Create a chart or graph of information found in a census.
- Make a list of five different ways census information is used.
- Provide five facts from the last census, including current U.S. population.
- Write the story of your birth. Include dates and times. For adopted or foster kids, write the story of when you first arrived in your current home.

Chapter Two

Concluding the Project

Questions to consider

- Were students able to locate information on their family?
- Do ancestors come from other countries?
- Do students understand all the uses of census information?
- Do students understand value of family information?

Activities

- Develop a family tree model for as far back as you can go.
- Create an "All-About-Me" poster telling your birthday and significant family members in your family.
- Write an advertisement convincing adults to participate in census information.
- Write a list of steps to follow when investigating your family's history.
- Act out a famous family story for your classmates.
- Provide a list of interesting facts and events that occurred on your birth date, include pictures.

Resources

Books

Climbing Your Family Tree: On-line and Off-line Genealogy for Kids by Ira Wolfman; Workman Publishing Company (October 2002)

Through the Eyes of Your Ancestors: A Step-by-Step Guide to Uncovering Your Family's History by Maureen Taylor; Houghton Mifflin Co. (March 1999)

Your Guide to the Federal Census: For Genealogists, Researchers, and Family Historians by Kathleen W. Hinckley; Betterway Pub (April 2002). I have included this book as background for teachers. It will enable you to answer the many questions students may have about using census information.

Internet Sites

http://www.rootsweb.com/~wgwkids/—Great site including worksheets and getting started tips. The site is written for kids under age 18.

http://www.genealogytoday.com/junior/—Another fantastic site written specifically for kids including games and activities to study genealogy.

http://www.census.gov/dmd/www/teachers.html—This site provides resources and materials for teachers to use in the classroom regarding census information.

Family Tree

All About Me

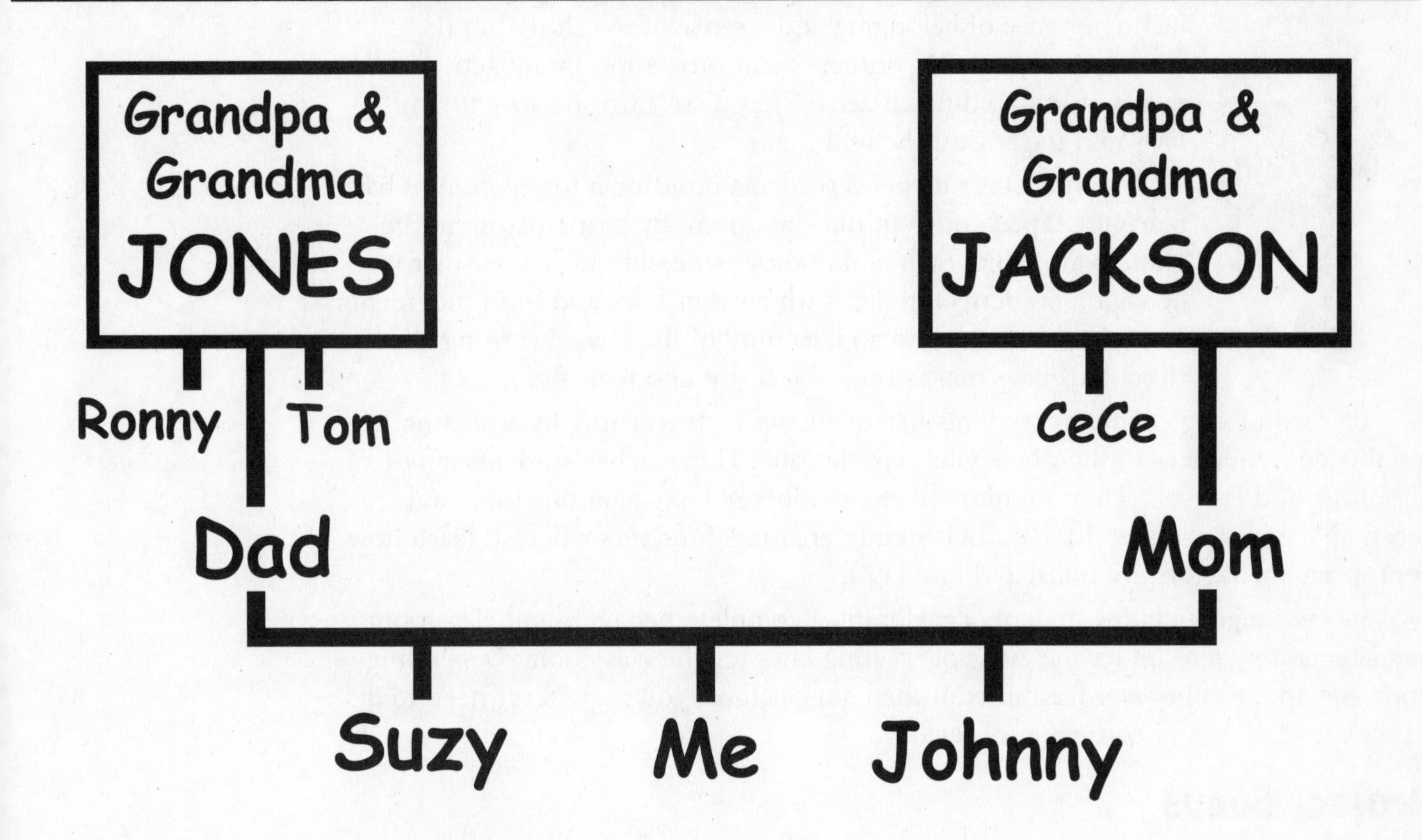

—**My Birthday:** July 4, 1995.
My birthday shares the same day with Independence Day.

—**My Favorite Food:** Macaroni and Cheese.
My grandma says it's her's too.

—**My Favorite Sport:** Just like my Dad, basketball is my favorite sport.

A sample family tree with noted traits.

Chapter Two

Social Studies
Grades 3-5

Law Abiding Citizens

Raising responsible, productive, law-abiding citizens is the goal of any basic education. Part of being a constructive person is feeling a sense of belonging and a sense of worth within the world. As part of this, students learn the appropriate actions and needs of law-abiding citizens. They also learn the structure of laws and the reasons behind them.

The first stage involves students building a foundation of basic behavior expectations in the classroom. By participating in the choices available, each child takes ownership over their own behavior. Students interact with current laws and form judgments as to the feasibility and applicability of the laws. Learning about ridiculous laws makes the subject fun and realistic.

The second stage contains students personalizing their learning by assigning reasonable consequences to their behavior expectations. This teaches students about reasoning and fairness. They are more likely to adhere to expectations they find acceptable and those they have participated in creating. Students will also learn how appropriate behaviors are translated into laws.

The last stage includes students developing a complete behavior and classroom management system under the guise of creating laws for the classroom. Depending upon the applicability and feasibility of their suggestions, you may never need to discuss expectations and consequences again.

Project Focus

As new members of your classroom democracy, create the laws and consequences for breaking these laws. All aspects of the laws must be reasonable and complete.

Why this project is valuable

Raising productive, law-abiding citizens is the goal of a well-rounded education. By affording students the opportunity to establish laws and consequences relevant to their lives, we enable them to self-discipline and monitor their own behavior.

Students will learn

- Responsibility for their actions
- How to solve a problem
- Developing a useable plan
- Reasonable options
- Nature of laws
- Purpose of laws
- Structure of laws
- Need for laws
- Consequences of breaking laws
- History of laws

Vocabulary

- Democracy
- Bill
- Law
- Congress
- Government
- Legislation

What kids should know before beginning this project

Students should have some knowledge of

- Democracy
- Laws
- Citizens
- Consequences

Beginning the Project

Brainstorm and discuss

- Purpose of laws.
- What happens when laws are broken?
- Democracy

Questions to consider

- What are core ideas to be learned?
- Do students understand the nature of a democracy?
- Do you have rules (laws) already established in classroom?
- Do any need to be changed?
- Do you have established consequences for behavior in classroom?

Chapter Two

Activities

- Create project journal.
- Work together to create a definition of democracy in your classroom.
- Make a list of five rules in your home.
- Compare your list to a friend's list. Tell what is the same and different.
- Investigate and report on the process of making an idea into a law.
- Draw a flow chart of steps for taking a bill to Congress to make a law.

Developing the Project

Questions to consider

- What are the basic guidelines for the laws? Life, liberty, and the pursuit of happiness? Safety, harmony, and equal education for all?
- How detailed will the laws be?
- How will the laws be written? Do not…, Thou shall not…., Honor, adhere, etc….

Activities

- Create a list of basic guidelines for behavior in classroom.
- Change each of these expectations into a law—How should it be written?
- Designate a consequence for each law—What happens when you break the law?
- Find a law you think is silly. Explain why you think it is silly.
- Choose a school rule (law) and explain in your journal why this law is important.
- Design a blank "ticket" for officers (students) to fill out and turn into the judge (teacher). Each ticket design should include space for the date, time, infraction, summary of incident and any other aspect deemed relevant in your classroom.
- Propose an enforcement plan. How will the laws be enforced? How can students participate in the enforcement without intruding on learning time?

Concluding the Project

Questions to consider

- Are these laws really going to be used?
- Are the laws feasible?
- Do students understand what each one means?
- Are consequences realistic, fair, and humane?

Activities

- Propose a "law and order" scenario for your classroom. Detail common laws, consequences, and methods for enforcing laws.
- Design signs to post in the classroom clearly depicting and reminding students of the laws for the class. Include ideas on where to post signs.
- Create basic "school laws" that would pertain to all students and the school. Take the proposal to the principal.
- Role play a student breaking the law and what happens after including consequences and enforcement.

Resources

Invite a police officer or lawyer to talk to the class about laws relevant to kids such as littering, jaywalking, etc.

Books

You May Not Tie an Alligator to a Fire Hydrant: 101 Real Dumb Laws by Jeff Koon and Andy Powell; Free Press (June 2002) Probably too hard for beginning readers but can be read aloud.

Loony Laws and Silly Statutes by Sheryl Lindsell-Roberts; Sterling Publications (August 1996) Again, needs to be read aloud.

Every Kid's Guide to Laws That Relate to Kids in the Community (Living Skills) by Joy W. Berry; Childrens Press (February 1988)

Internet Sites

http://www.congressforkids.net/makinglaws.htm—Another great government site that includes information on how a bill becomes a law, also trivia game and background information.

http://www.lawforkids.org/—This site is slated for kids in Arizona and Arizona laws but has tons of fun activities and information for kids in general. Animated clips tell what happens when kids break the law, links provide access to games relating to laws, and much more.

http://dirs.educationworld.net/cat/473447—Follow the links to read about laws specifically for kids in California or to read about a second grade class that "used the legislative process to make the ladybug the Official Bug of the State of Massachusetts."

http://www.education-world.com/a_lesson/lesson274.shtml—This site provides thoughts and ideas behind establishing classroom rules. It also provides links and ideas to use in the classroom.

Law Abiding Citizens

WARNING!

FINE FOR:	# OF TOKENS
Personal artifacts on floor	5
Trash around desk	5
Ugly comments to classmates	10
Disruption to class	10

Be Smart! Don't Break The Law!

A sample chart of 'laws' for the classroom.

Social Studies
Grades 3-5

Reinvestigating History

This project has very strong implications about altering history and the future. Kids will run away with it. Understanding history and how it shapes the future allows students the ability to recognize the importance of all aspects of life as it unfolds. Deep discussions could ensue with this project. It is perfect for this age group.

In the beginning stage of this project, students learn to build the foundation of knowledge for the historical event that they are studying. They become versed about the event, the key players in the event and major factors that contributed to the overall circumstance. Each of these learning experiences will develop their knowledge of the historical significance of the event.

During the developing stage, students further their understanding of the event by specifying specific details and relating information into a cause and effect format. They build a working vocabulary and a timeline in history.

In the concluding stage, children will test their reasoning, problem solving, critical thinking, and creativity. Each aspect of the final product is based on supposition alone. The depth, complexity, and reasonableness of their suppositions will depend upon their ability level. For less accelerated students, suggestions may be necessary.

Project Focus

Your time machine has taken you back to a different period in time. Create a plan to alter history by changing the outcome of a certain significant historical event. For example: You have arrived in Texas in the mid-1830s. Create a plan to beat Santa Anna at the Alamo and win the battle.

Why this project is valuable

Learning about history helps students understand their world. It also gives them perspective into their world.

Chapter Two

Students will learn

- How to solve a problem
- Developing a useable plan
- Significance of historical event
- Significance of altering historical events
- Idiomatic language—such as "Hindsight is 20/20."
- Point of view
- Perspective

Vocabulary

The vocabulary will differ with each historical event. Pick the key words students will benefit from retaining after completing the project or those that are specific and valuable to the historical event.

- Various idioms—discuss significance of these idioms and time period from when the idioms were born.

What kids should know before beginning this project

Students should have some knowledge of

- How to perform basic research functions.
- How to work together.
- How to read a map.

Teacher's Tip

To develop an understanding of point of view read the following two books and discuss prior to beginning or as an opening activity:

The Pilgrims of Plimoth by Marcia Sewall—this book gives an account of daily activity in the Plymouth Colony and some brief history.

People of the Breaking Day by Marcia Sewall—details the same information from the Indians point of view.

Beginning the Project

Brainstorm and discuss

- The basis of the event—Will everyone work on the same event (due to curriculum needs) or may students pick their own event?
- Structure of the government at the time.
- Basic lifestyle of the time.
- History shaping the future.
- Relevance of historical events in relation to other historical events.
- Important people in history.

Questions to consider

- What is the significance of the event?
- What happened?
- Who are the key players?
- What is the end result?
- Why would it have been valuable to change the outcome?
- From what perspective are you studying the event?
- From who's point of view is the subject being studied?

Activities

- Create project journal.
- Create a cause and effect t-chart showing key events that occurred.
- Map out the events and movement of key players in event—copy a map of the time and write directly on it.
- Write the history of the event in your own words.
- Write a pretend interview of one of the main characters of the event—i.e.—Sam Houston, George Washington etc.
- Roleplay the interview on video.
- Pretend you are one of the key players in the event; write a sample diary entry describing your thoughts about the event.

Chapter Two

Developing the Project

Questions to consider

- Why would you want a different outcome?
- How could altering the event change the future?
- Would the change be good/bad/indifferent?
- What were the thoughts and feelings of the key players?
- How would you (the student) feel in that situation?

Activities

- Write a play depicting specific aspects or key events.
- Roleplay possible conversations between key players in the event.
- Design clothing for a key player depicting the time period.
- Draw a map showing movement of soldiers or other significant people's movements through event.
- Create a film strip or story board depicting the significant events in a timeline manner.
- Create a diorama about the actual event.
- Create a timeline displaying the minor events that led to the major event.

Concluding the Project

Questions to consider

- What is the final result?
- What significance could the changed outcome have had on the future?
- Why is it important to not really be able to change the past?
- What can be learned from this event?
- What generalizations from this event might be given to help in planning for the future?

Teacher's Tip

Have students focus on ONE aspect of the event. Trying to investigate too many pieces might create an overwhelming feeling and make the task too difficult.

Activities

- Draw a map keeping to the landforms and structures of the time outlining the new route or plan.
- Rewrite the event for history books, detailing the "new" outcome based on the different choices, decisions made.
- Create a t-chart representing what really happened/what could have happened (knowing what we know now—hindsight).
- Create a mural showing new historical perspective.
- Create a new film strip or story board telling the new history.
- Create a poster representing the possible changes in history if the event could have been altered—i.e.—what would have happened to San Jacinto if we had won at the Alamo?
- The battle cry of the men that won at San Jacinto was "Remember the Alamo!" Write a new battle cry or slogan for your event.

Resources

Books

These will vary depending on the historical event being studied—check with the library.

Television

The discovery channel has a new show titled *Unsolved History*. This show investigates physical evidence from history. It looks at the evidence to recreate the historical perspective of certain events. This would bea great show to watch to give students perspective on looking at things differently.

Internet Sites

For Kids

www.ask.com—This is the Ask Jeeves search site that is safe for kids.

http://sunsite.berkeley.edu/KidsClick!/—Search engine for kids.

For Teachers—Background and resources for teaching historical events.

www.ncss.org—National council for the Social Studies.

www.history.org/nche—National Council for History Education.

Reinvestigating History

Winning at the Alamo!

Historical Perspective	Hindsight Perspective
Sam Houston didn't know that Santa Anna's army was moving so quickly.	**Sam Houston** could have sent spies to track Santa Anna's army.
All of **Fannin's Army** was killed before getting to the Alamo.	**More troops** could have been sent to help Fannin. Then Fannin and his troops could have fought at the Alamo.
Susanna—wife of Houston's sidekick endured the fighting at the Alamo. Later, she told her story.	If **Susanna** had decided to leave before Santa Anna reached the Alamo, we may never know the truth about what happened.

A sample before/after scenario.

Social Studies
Grades 4-6

Family Vacation

What fun vacation planning is! Everybody loves to dream of faraway destinations. This project provides an avenue for students to dream while learning valuable skills and concepts. The travel aspect lends itself to all types of social studies objectives. By specifying the need for educational destinations, the number of possible visits is refined. You may remove or adjust this stipulation to fit your needs.

During the beginning phase, each student develops destination spots or possible educational outlets and adventures. Many students will be surprised at the cost of travel. Use this opportunity to teach mathematic objectives. The class will enjoy dreaming about exciting places to go and the dreaming helps to release their creative juices.

In the developing stage, the class personalizes their learning by deciding food selections, proposing wardrobe choices, and finalizing destination plans. Each step will heighten student excitement. As their excitement grows, students will engage themselves in their learning. For those students that never experience the excitement of vacation planning, they'll still learn valuable social studies objectives.

In the concluding stage, children propose travel destinations and itinerary suggestions for their fictitious vacation. Students should be certain to include all pertinent details and adhere to specific predetermined criteria.

Project Focus

You are responsible for planning your family vacation this year. Your parents want it to be an educational vacation that everyone will enjoy. Plan the full itinerary for your trip.

Why this project is valuable

Not only will students learn many valuable planning and problem solving skills, but they will also learn to develop focus by honing-in on specific vacation aspects.

Chapter Two

Students will learn

- Developing a useable plan
- Using timelines
- Backing out time
- Reading a map
- Modes of travel
- Various cities and/or countries
- Various cultures or heritage dynamics
- Complexities of planning a vacation
- Purpose of travel agents

Vocabulary

- Itinerary
- Destination

What kids should know before beginning this project

Students should have some knowledge of

- Time, days, months, seasons, etc.
- Reading a map
- Budgetary guidelines—if any

Beginning the Project

Brainstorm and discuss

- What constitutes a vacation—set some parameters like going to another place for an extended period of time for relaxation?
- Have students been on a "formal" vacation?
- Favorite types or spots for vacation.
- Favorite way to travel—boat, plane, train, automobile.
- Set the parameters such as how long, how much money can be spent etc.
- Let it be a dream vacation. They can do anything they want, but remind them the more involved the more work and details.

Teacher's Tip

Dictate a specific state or country for the vacation depending on your curriculum needs.

Questions to consider

- What does your family like to do?
- Do you like warm or cold weather?
- Are there specific activities you want to do on your vacation?
- Is there a theme to your vacation?
- How many members of your family will be going?
- How long will the vacation be?
- What educational opportunities will there be?
- What amenities will be part of the vacation?

Activities

- Create project journal.
- Record in journal feelings about a vacation and expectations for learning along the way.
- Decide educational experiences and justify their value in your journal.
- Create a list of all places to travel.
- Create a list of all expenditures.
- Draw a picture of your trip.
- Write an interview for a travel agent.
- Contact a local travel agent and perform interview.
- Create a display or summary of a travel agent's job.

Developing the Project

Questions to consider

- Do students understand how to plan ahead—if you have to be somewhere at 3:00 p.m., what time should you leave, how long will it take to get there?, etc.
- Are the choices students make feasible?
- Are there any special requirements with their choices such as passports?
- Does anyone in the family have special needs?
- How will you get to where you are going?
- Where will you stay once you are there?

Teacher's Tip

Real vacation planning must include incidentals and minor issues. You many not want to make kids get overly involved in minor aspects. This could deter from the overall project. Some students may need the time of vacation or travel destination shortened or lengthened depending on needs.

Chapter Two

Activities

- Create a comparison chart between two favorite destinations to help decide which one will be the final. Use a t-chart listing pros/cons.
- Write an individual "need-to-do" list for each member of the family.
- Design a plan of what to do days before leaving on the trip—present in any format.
- Investigate and record climate and weather patterns for area visiting.
- Suggest necessary wardrobe for trip.
- Study the culture or dynamics of the people, decide on food choices.
- Create a diagram, picture, display of some sort showing culture or dynamics of people in area.
- Create a display showing various lodging choices.

Concluding the Project

Questions to consider

- What time and budgetary guidelines were set at the beginning?
- What resources are available?
- What educational activities or sites are available?

Activities

- Create in paper, digital, or poster format a complete timeline itinerary for trip.
- Draw a map indicating the routes to be taken during trip designating distance traveled from point to point. Label with important landmarks.
- Create a complete vacation presentation using classroom materials or technology. Justify the learning experiences at each vacation spot. Detail who, what, and when of every stage of the vacation.
- Compare two travel destinations giving the pros/cons of each destination. Try to create equal situations. For instance, comparing Disney Land and the local zoo is not equal.
- Design a brochure advertising your favorite vacation destination, including educational opportunities offered.

Resources

Consult a local travel agent. Ask them to come speak to the class concerning things to consider when planning a vacation.

Books

Smart Vacations: The Traveler's Guide to Learning Adventures Abroad by Council on International Educational Exchange; St. Martin's Press (March 1993) There are travel books and guides for just about anywhere. Once students locate specific destinations, they may need to check out specific books at the library.

Internet Sites

www.insidetrips.com—This site provides excellent tips for planning a vacation.

http://www.familytravelguides.com/tips.html—Another site appropriate for kids with plenty of travel tips.

Family Vacation

Dave's Vacation Itinerary

Destination: Disney World, Orlando, Florida

Departure Date: March 15, 2003

Arrival Date: March 18, 2003

Educational Stops:
Shreveport, Louisiana
Jackson, Mississippi
Montgomery, Alabama
Tallahassee, Florida

At each of these stops, we intend on shopping at the...

A sample itinerary and stopping points.

Chapter Three

Language Arts Projects

Grades 2-3

Comic Relief

Comic books and comic strips were once a large contributor to many students' reading lives. Over time the trend has faded, yet comic strips are still very much a part of newspapers and magazines. Learning about this genre exposes students to an avenue for creativity and enables students to understand the structure and history behind this huge genre of art.

In the first stage, students will develop an understanding of the many different types and forms of comic strips. Each activity will build the foundation of the structure and use of comic strips. By investigating common traits and elements of comic strips, students will begin to develop trends and patterns used in this genre.

During the second stage, your class will personalize the information by deciding on factors that will be used in developing their own comic strip. They will learn story elements and character traits from previous comics and relate that information to their own comic strip.

For the last stage, students need to demonstrate their own learning by fully developing their own comic strip. Students may publish their strip through a variety of avenues depending upon resources available to them. The end result is a demonstration showing they understood the structure and elements in their own comic strip.

Project Focus

Your newspaper needs a new comic strip. Create one to be published in your local or school newspaper.

Why this project is valuable

Students will learn

- Responsibility for their actions
- How to solve a problem
- Developing a useable plan
- The history of comic strips
- The purpose of comic strips
- How to develop a comic strip
- Sequencing
- Main Idea
- Facts and Details
- Dialogue
- Character Development

Vocabulary

- Comic Strips
- Superheroes
- Character—as part of a story
- Dialogue
- Sequence

What kids should know before beginning this project

Students should have some knowledge of

- What a comic strip is

Different types of comics

- Superheroes
- Political
- Family

Other objectives that will be taught prior to project beginning such as:

- Sequencing
- Main idea
- Facts and details
- Historical facts as part of a story line
- Any other aspect of the curriculum that could be tied to this project

Beginning the Project

Brainstorm and discuss

- What is a comic strip?
- What is the purpose of comics? To entertain, send a message, represent ideas
- What do they look like? Colorful, black and white, sketched, computer generated?
- How are they organized? One frame, multiple frames

Chapter Three

Questions to consider

- Is there a specific objective to be taught through the comic such as?
- Sequencing—first, next, last.
- Main idea—overriding theme or idea.
- Facts and Details
- Character development
- Dialogue
- Should the characters and story line tie to a specific subject?
- Could the student's comic be similar to another one or must it be completely original?

Activities

- Create a project journal.
- Have students bring in examples of comics from newspaper, parents collectables, or books—discuss.
- Draw a web with comics in the middle and all words associated with them on the outside, include in journal.
- Read several comics together—have students choose their favorite and explain why it is their favorite in their journal.
- Plot the sequence of events contained in their favorite comic.
- Explain the main idea in their comic.
- Compare/Contrast two comics from a local newspaper, book source etc.
- Find common elements in all comics—make a list or draw distinctions as class, group or individual.
- Make a 3-D figure of a comic hero out of paper, boxes, plaster of paris, paper mache.

Developing the Project

Questions to consider

- Who will the characters be?
- What will the story line be?
- Develop the story line by providing a main idea and sequence of events.
- What is the purpose—entertain, send a message?
- What format will be used for the comic—computer generated, hand drawn, color, black and white?

Activities

Conduct an interview with an author of a comic strip. Some questions to ask:

- How do you decide your characters?
- From what do you develop your story line?
- How long does it take to develop each strip?
- How long have you been doing comic strips?
- Is there a particular message for your strip?
- Do you have any advice for someone wanting to make their own comic strip?
- What should I do? What should I do?
- Design the characters to be included in your comic strip—what actions will they take?
- Plot the sequence of events to occur in the comic strip on a time line.
- Display daily cartoons and have students write in their journal specific skills being taught. For example:
  - What is happening in the comic?
  - What is the setting of the comic?
  - What is the message of the comic?
  - List all the verbs, nouns, adjectives etc.
  - Design the setting and look of the cartoon.

Concluding the Project

Questions to consider

- How will the information be presented? Paper, computer generated, etc?
- Will students work from a sample or generate something completely original?
- How lengthy will the project be?

Activities

- Create a story board comic strip to display in the classroom, with a paper version for the newspaper.
- Create a completely original multi-frame comic to publish directly to the newspaper.
- Create a slideshow presentation of scanned comic frames to create a digital comic strip, send hardcopy to newspaper.
- Create a diorama or other 3-D representation of the comic strip, send paper version to newspaper.
- Conduct a play or short skit acting out comic strip, send paper version to newspaper.

Chapter Three

Resources

Books

How to Draw Comics the Marvel Way by Stan Lee; Fireside (September 1984)—This book could be bit difficult for early readers but is an excellent resource and could be read aloud in pieces or completely to all.

Calvin and Hobbes Sunday Pages 1985-1995 by Bill Watterson; Andrews McMeel Pub (September 2001)

Any other comic book.

Internet Sites

http://www.comic-art.com/history.htm - This page offers fantastic background on the history of comics.

http://www.mogozuzu.com/comics.htm - More history on comics.

Teacher's Tip

Caution! There are comic books with inappropriate content for school-age children. Be careful when selecting resources.

Comic Relief

Safety Sally had to think quick when the rain came. Good thing she thought to go to the neighbor's house during a storm.

Safety Sally says...

Be smart during a storm!

A sample comic with a public message.

Chapter Three

Language Arts
Grades 2-3

Guide Word Thesaurus

Learning to use common resources that are available is a valuable tool for all students. By understanding the structure of dictionaries and the purposes of a thesaurus, students are better able to access information when it is needed.

First, students find the purpose and use of guide words in a dictionary. They will also learn the differences and similarities of dictionaries and thesauri. By investigating common roots, prefixes and suffixes, students will gain an understanding of the structure of words.

Next, they will personalize the information by choosing their own words to use. Then, they use the meaning of these words and how to graphically represent these words.

Last, students are required to apply their learning by developing their own picture thesaurus. After using the words repeatedly, they'll learn valuable word study skills and digest what is learned. In the future, the students will develop a working knowledge of how to use these resources for other learning opportunities.

Project Focus

You keep hearing words that just don't make sense. You think you know what they mean but you aren't sure. Using a page from an elementary dictionary, create a picture and word thesaurus.

Why this project is valuable

All students will use a dictionary and thesaurus from time to time. A key element to understanding these books is to use the guide words at the top of the page. Learning to use these valuable resources will enable students to learn in on their own.

Students will learn

- Responsibility for their actions
- How to solve a problem
- Developing a useable plan
- Guide Words in a dictionary
- Organization of words
- Alphabetical Order
- Synonyms
- Picture Representations
- Vocabulary Development

Teacher's Tip

For struggling readers, picking a page from a dictionary and working strictly from it might be too difficult—for these students let them choose some of their favorite words and proceed as the rest do.

Vocabulary

- Dictionary—must understand not all dictionaries are the same
- Thesaurus
- Guide Words
- Vocabulary words
- Synonyms

What kids should know before beginning this project

Students should have some knowledge of

- The organization of words in a dictionary.
- How dictionaries differ.
- The different information in a dictionary.
- The words and organization of a thesaurus.
- Where the guide words are located.

Beginning the Project

Brainstorm and discuss

- Why are guide words in a dictionary?
- Why are dictionaries useful?
- What information is contained in a dictionary?
- What is a synonym?
- Why is a synonym useful?

Questions to consider

- How are words organized in a dictionary? In a thesaurus?
- What makes up words? Sounds, roots, prefixes, suffixes.
- Are there common prefixes and suffixes to look for?
- What are the meanings of common roots, prefixes, suffixes?
- Who will decide which page to use from the dictionary—student or teacher?
- Will the page be done individually or as a group? Base this on kids' abilities.
- How many pictures per definition, is there a limit or minimum?

Chapter Three

Activities

- Create project journal
- Create a chart labeling common roots, prefixes and suffixes.
  This could be done as a class as a reference for everyone or if more time allows, each student or group can create their own, including pictures.
- Make a list of other words using these common prefixes and suffixes.
- Explain rational for choosing specific page from dictionary in journal.
- Brainstorm possible pictures to include in picture thesaurus.

Developing the Project

Questions to consider

- What format will be used to present this picture thesaurus page?
  - Multimedia slideshow
  - Paper book
  - Poster
- Will pictures be of words from dictionary or synonyms (thesaurus words)?
- Where will students look for definitions and synonyms?
- Could antonyms be used if synonyms are too easy or too hard?

Activities

- Write all words from dictionary page in journal and define.
- Find synonyms for each word in thesaurus, off internet, talking to another adult.
- Draw a picture representing each definition.
- Write a sentence with the dictionary word and then put a synonym in its place.
  - Does the synonym change the construction of the sentence?
- Construct a quiz asking questions to determine if a certain word would fit within your guide words on your sample dictionary page .
  - Give the test to a classmate and grade.
  - Record results and conclusions in journal.
- Have each student pick their favorite word and create a costume depicting definition.
- Hold a "fashion show" for all the words.

Concluding the Project

Questions to consider

- How will the information be organized?
- One definition per page or multiple?
- How will pictures be made?
    - Photographs
    - Digital pictures
    - Magazine cut-outs
    - Drawing
    - Download from computer
    - Combination of all?

Activities

- Create picture thesaurus using multimedia, paper, crayons, markers, etc.—make sure the words are from the dictionary page chosen or an established word list.
- Develop organizational chart of roots, prefixes, and suffixes for individual words.
- Using graph paper or software program make a word search for your picture thesaurus page.
- Write the history of the dictionary and thesaurus in your own words.
- Recommend a list of your favorite words to a student in a lower grade—provide accurate definitions and uses for the words—share with other classes.

Teacher's Tip

This is a great time to have extra word puzzles around the classroom for kids to interact with words in many formats.

Chapter Three

Resources

Books

The Professor and the Madman: A Tale of Murder, Insanity, and the Making of the Oxford English Dictionary by Simon Winchester; G K Hall & Co. (April 1999)—This is not for kids but provides an outstanding background for teachers.

Multiple dictionaries and thesauri should be used to demonstrate differences and similarities. Don't forget to mention dictionaries on PDA's and those used in word processing programs.

Internet

www.wordcentral.com—This site is for students and teachers.

http://www.enchantedlearning.com/Dictionary.htm—This online dictionary provides pictures with the definition—Great site!

http://www.funbrain.com—The many games on this site make for great learning opportunities. Be sure to monitor students though. Some students could be easily distracted.

Word Guide Thesaurus

A Study of Adjectives

Dark...
dim
cloudy
dull

Silly...
goofy
funny
crazy

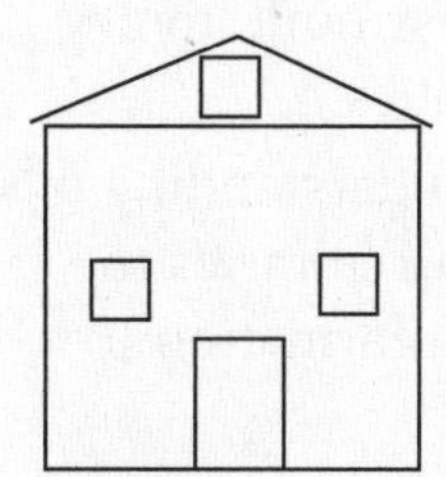

Big...
huge
gigantic
large

Sunny...
warm
happy
bright

Cute...
small
cuddly
pretty

Slow...
creep
crawl
lazy

This is a sample page from a picture thesaurus.

Chapter Three

Language Arts
Grades 3-5

Classroom Library

Libraries can become overwhelming for small children. By studying the structure of a library, the job of a librarian, and the types of resources available, students develop an understanding of libraries. By examining each individual factor in a library, learning about libraries is easy. Students develop a big picture idea that is then used on a small scale in their own classroom.

In the beginning stage of the project, each child seeks out various genres of literature available as well as examples within each genre. Students will learn about the job of a librarian and the many tasks required within that job description.

During the developing stage, the class has to think about likes and dislikes of classmates with regard to various genres. This information will be used to determine necessary amounts of books for each genre to be placed in their classroom library. Students learn valuable survey and response skills from these activities.

Lastly, in the concluding stage, students must assimilate their understanding of the various genres, their classmates' preferences, and the structure of libraries to develop a plan for a fictitious classroom library. They can use real information to determine the outcome of their plan.

Project Focus

The teacher wants to establish a classroom library according to her students likes and wants. Survey the class to determine what type of books should be in the library. Then decide how many of each type of book should be available.

Why this project is valuable

Students will learn

- Responsibility for their actions
- How to solve a problem
- Developing a useable plan
- How to conduct a survey
- How to compile information
- How to create a graph
- Organization
- Using data to determine outcomes
- The Dewey Decimal System

Vocabulary

Different genres—mystery, historical fiction, fiction, nonfiction, biography, plays, etc.

Dewey Decimal System—as an organizational tool.

What kids should know before beginning this project

Students should have some knowledge of

- Definition of different genres.
- Personal like and dislikes when reading.
- Purpose of a library.
- Where the school's library is located.

Beginning the Project

Brainstorm and discuss

- Libraries.
- Various genres of literature.
- Likes and dislikes of classmates.
- How libraries are organized.
- What a librarian's job is.
- Various types of librarians—some do more than just deal with books, many deal with technology and computer labs as well.

Questions to consider

- How much space will be available for this library?
- Approximately how many books will be included?
- Where will the books come from?
- Who will have final say in the overall organization of the library?
- Create a challenge situation—the best really gets used!

Chapter Three

Activities

- Create project journal.
- Journal personal likes and dislikes when reading.
- Categorize already read books into genres.
- Analyze data and look for personal reading patterns, i.e.—does the student only read fiction books?
- Read aloud short books or pieces of books from every genre to expose kids to each type.
- Invite your school's librarian in to discuss job requirements.
- Investigate necessary educational requirements to become a librarian.
- Explain in your project journal your understanding of the Dewey Decimal System.
- Conduct research on how libraries developed and why. Report findings.

Developing the Project

Questions to consider

- What impression do the kids have about each genre?
- Have they been exposed to all genres?
- What system is used to track books in your school's library?
- Do students understand the Dewey Decimal System?

Activities

- Take a tour of your school library - Have librarian explain organization of this specific library.
- Write a survey to determine class's favorite genres.
- Perform survey.
- Graph results from survey.
- Using survey results, graph all genres to be included in library.
- Using survey result, graph number of books per genre to be included in library.
- Develop possible ways to track checking in and out of books.
- Explain if the Dewey Decimal System would work in your classroom library.
  - Detail why or why not.

Concluding the Project

Questions to consider

- Will the proposals really be used or is this for a fictitious library?
- How will the proposal be presented?
- What design elements will be included in the proposal?
- What tracking system will be used?

Activities

- Create a proposal for organization of books, including how to keep track of them—check in/out procedure.
- Draw a design of where to place books, allowing extra room for favorite books.
- Provide a written survey analysis of which books are liked the most and how many to include.
- Create a graph displaying favorite books and how many to include of each.
- Create a display (written or otherwise) depicting all qualifications and job duties of a librarian.
- Propose a list of books to be read from each genre.
- Write a letter to a student just learning to read about the benefits and values of a well-organized library.
- Make a T-Chart showing pros/cons of being a librarian.

Teacher's Tip

As an extension, require students to document their reading of one book from each genre in the next grading period or semester.

Chapter Three

Resources

Books

A Day in the Life of a Librarian (The Kids Career Library) by Liza N. Burby; Powerkids Pr 1999

Librarians by Dee Ready; Bridgestone Books 1998

Learning about Books and Libraries: A Goldmine of Games by Carol Lee and Janet Langford; Highsmith Press (April 2000)—This book is a teacher resource guide and would be valuable in providing additional information.

Internet Sites

http://www.mtsu.edu/~vvesper/dewey.html—Written for students, this site is filled with information and activities to engage students in learning about the Dewey Decimal System.

http://www.mte.asd103.org/library/dewey/deweystory.htm—This site will provide sufficient explanation for the teacher and provides a link at the bottom of the page for lesson ideas that will further your thinking and get those juices flowing!

Classroom Library

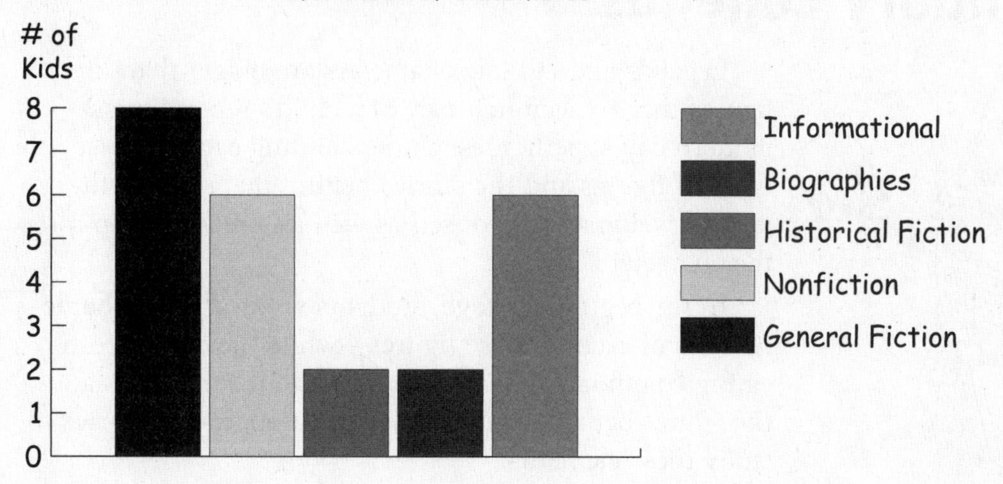

Our Proposal: <<# of books/genres>>
Total # of books = 130

Number of bookshelves: 5

Number of books per shelf: 20-30

Total Books: 100-150

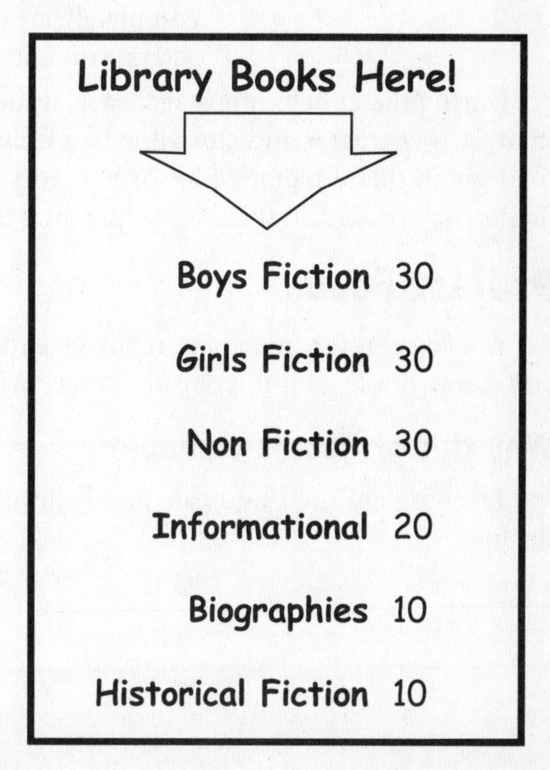

Sample Project

Chapter Three

Language Arts
Grades 3-5

Legendary Legends

Legends and tall tale characters are bigger than life figures that are as much part of a child's repertoire as modern day superheroes. Understanding each of these types of figures and the stories behind them helps students develop a sense of self as well as a relationship to the characters.

In the beginning stage, students will learn the characteristics of each of these figures—while they are investigating common elements pertaining to these figures and the stories behind them. They will compare real stories to study these elements.

Students have to personalize their studies by developing characters and story elements specific to their tastes during the developing stage. Then they'll develop their character and story line based on previous stories or by completely creating new ideas. By illustrating their characters, students will visualize the outcome of their story.

During the concluding stage, each student exhibits their learning by showing their own character and story line in a fully produced story. Each pupil will have a final say in the outcome of his or her story and character traits. Specific literary elements may be tied to the story depending upon curriculum needs and goals.

Project Focus

You've been reading about many legends lately. You feel certain you or someone you know is a legend in your own time. Write your story.

Why this project is valuable

Teaching children to dream and believe in themselves is the best ammunition for life.

Students will learn

- Responsibility for their actions
- How to solve a problem
- Developing a useable plan
- Learning to dream
- Believing in self
- Structure of stories
- Story development
- Tall tales, folklore and legends
- Story plot
- Story characters
- Story setting
- Common elements of legends and tall tales

Vocabulary

- Nature
- Superhuman
- Legends
- Tall tales
- Folklore
- Nature
- Story elements—plot, setting, characters, climax.

What kids should know before beginning this project

Students should have some knowledge of

- Various genres of literature, including tall tales, legends and folklore.
- Story elements—plot, setting, characters, climax etc.

Teacher's Tip

Steven Kellogg is the author of many adaptations pertaining to tall tales. I suggest collecting many examples from various authors. Exposure to a variety of stories will help students release their imaginations.

Chapter Three

Beginning the Project

Brainstorm and discuss

- Legends, folklore, and tall tales.
- Historical figures as legendary figures.
- Superhuman acts.
- Nature as a part of legendary tales.

Questions to consider

- Will there be a model for your legendary tale?
- What key aspect makes them a legend?
- Will each tale include superhuman powers or nature?
- If no superhuman powers or nature, what makes them a legend?
- Will aspects of nature be a part of the story?

Activities

- Create project journal.
- Read aloud *The Legend of the Paintbrush* and/or *The Legend of the Bluebonnet* by Tomie dePaola. Discuss story structure, plot, setting, characters, and legend including nature behind story.
- Read aloud or silently tall tale stories of Davy Crockett, George Washington, Pecos Bill, Johnny Appleseed, Paul Bunyan, or any other tall tale. Discuss and write what makes each a tall tale.
- Read two versions of Pecos Bill—chart similarities and differences.
- Ask a folklore expert in your area to discuss with class nature and origin of such stories.
- Make a list of common elements in legends and tall tales.
- Plot the story elements of your favorite legendary tale.

Developing the Project

Questions to consider

- What makes you a legend?
- How will the students develop their plot?
- Are they working together or alone?
- Can they use another story as a guide?

Activities

- Draw a picture of you or your main character.
- Create a story plot for your legendary story.
- Draw illustrations for your story.
- Write and perform a scene from your favorite tall tale.
- Compare your legendary story to another legendary story. Tell how they are alike and how they are different.
- Draw a time-line of events to occur in your story.

Concluding the Project

Questions to consider

- What multimedia is available to use?
- Do students have other story elements they are learning and need to include?
- Will they be performing or presenting their stories?

Activities

- Write a complete story including pictures, publish as a book.
- Provide a story map and outline for your legendary story.
- Write a script and act out your legendary story.
- Use multimedia to publish your story and present to a younger grade.
- Write a comparison chart for five different legendary heroes. Choose four main points to compare.

Chapter Three

Resources

Contact local universities, community colleges, historians, or any other expert on folklore, legends, and tall tales.

Books

American Tall Tales by Mary Pope Osborn; Knopf (September 1991)—This book contains many tall tales with background information - a fantastic beginning book to develop understanding of the topic.

American Tall Tales by Adrien Stoutenburg et al; Scott Foresman (October 1976) Again, a general collection—great for comparing to other books.

I Was Born about 10,000 Years Ago by Steven Kellogg; Mulberry Books Reprint (August, 1998)

Internet Sites

http://www.hasd.org/ges/talltale/talltale.htm—This site provides great background on tall tales in general.

http://www.americanfolklore.net/tt.html—There are many tall tales on this site, including stories specific to various regions of the country.

http://www.ga.k12.pa.us/academics/ls/4/la/4r/talltale/ttintro.htm—This site provides more specific lesson plan ideas for developing and using tall tales—it is a great teacher resource.

Legendary Legends

Everyone agreed that Dr. Joe was the best veterinarian in the Outback. When Mr. Bungle's kangaroos kept getting sick, the first person he called was Dr. Joe... and that's how Dr. Joe got springs in his legs.

Sample Legendary Character & Storyline.

Chapter Three

Language Arts
Grades 4-6

In Search of Bias

Bias is a natural outcome of different personalities and opinions. By understanding bias students are able to identify various positions relative to important issues in their lives. Each opportunity brings new light to the students' understanding of various topics. In realizing different viewpoints, students are able to place themselves outside the context of their world.

During the beginning stage of the project, students work to identify with their feelings and the feelings of people around them. They also carry out an understanding of how prevalent bias is in our society and the many forms in which it can be found.

Throughout the developing stage, students labor to form opinions about the nature of bias and the many forms of it. They also learn to evaluate bias and translate the evaluations to other opportunities. Students learn how to use bias to their benefit and discover when an article or show is biased.

In the concluding stage, students use their new knowledge to create variations on current issues and biased material. They learn to use specific words to slant materials in a different or opposite manner. By using current issues, students learn to relate to and form opinions about their world.

Teacher's Tip

For students who struggle with this concept:

- Provide the articles for them.
- Underline all words that contribute to the bias.
- Determine what the position of the article is—what is the bias—where are they (article writer, news reporter) are coming from.
- Use several other articles to determine same items—repeat until concept is internalized.

For higher-level students:

- Perform a market analysis of TV based commercials.
- How often is biased information presented?
- Is it just in commercials or is it in TV programs as well?
- If so, give examples, citing comments made and in what context they were made.

Project Focus

You find it impossible to believe the story you read in the newspaper this morning. It just can't be completely true. Using newspapers or magazines readily available identify bias within each piece. Rewrite the piece without bias or slanted the other way.

Why this project is valuable

There are always two sides to a story. Students must learn to understand perspective and bias in the news. Understanding bias helps youngsters form their own opinions. To form a true opinion, one must understand all sides of an issue. As future voters, students must learn to understand multiple sides to an issue.

Students will learn

- Responsibility for their actions
- How to solve a problem
- Developing a useable plan
- Nature of bias
- Nature of news
- Constructive writing
- Effects of bias on news

Vocabulary

- Bias
- Perspective
- Loaded Language

What kids should know before beginning this project

Students should have some knowledge of

- Newspapers and magazines
- Point of view
- Who, What, When, Where, Why, and How in news and articles

Chapter Three

Beginning the Project

Brainstorm and discuss

- Who writes articles in newspapers and magazines? Authorities on topics, in-house writers, freelance writers, politicians, reporters, etc.
- Why do they write these articles—what's the purpose? Money, job, opinion, change other's thinking.
- Who is the audience? Who reads the articles or watches that channel?
- What are the demographics of the audience?

Questions to consider

- How many types of news articles are there?—for example, informative, press releases, editorials, advertisements.
- When you read a newspaper or magazine article do you believe everything you read?
- What makes you believe or not believe?
- What facts support your belief?
- How can news or written articles make you feel?
- What types of words could be considered loaded language?

Activities

- Create project journal.
- Find an article that you disagree with and explain in your journal why you disagree with it.
- Search the library, Internet, ask professionals, etc. for facts to support the belief or disbelief.
- Identify and use an article that creates strong feelings.
  What are these feelings?
  What words or examples make you feel this way?
- Contact a local newspaper, TV station, or magazine, request demographic information of their audience.
- Analyze this information and provide a summary of their market.
- Watch their programming and see if it fits their market—is their bias intentional then?
- Journal findings and opinions.

Developing the Project

Questions to consider

- How do you identify bias? For example, loaded words, word emphasis, examples or consultants included in article.
- What are your feelings about bias? Good/Bad/Indifferent/Don't Know.
- Are there instances when bias should never be a part of an article, what are they?
- Are there instances when bias should be a part of an article, what are they?
- What is point of view?
- How relevant is it to learning about bias?

Activities

- Using a newspaper or magazine article chosen by the student, have students identify words that indicate bias.
- Read *The Three Little Pigs and The True Story of the Three Little Pigs* and discuss point of view of each. These stories are not too young for this age students. Students will be better able to focus on the differences and perspective when they are mostly familiar with the story line.
- Chart differences and similarities with T-Chart, Venn Diagram, etc.
- Explain in detail the point of view of each book.
- Students should pick a hot topic for their school, town, city, state or a national topic upon which there has been a great deal of media.
- Locate article representing opposing views on this topic.
- Compare the two articles—identify the bias in each.
- Decide on several topics (whatever number you choose) that should be biased for kids sakes, such as:
  - Smoking and kids
  - Underage drinking
- Make a should/shouldn't chart for topics that should be biased, in the students opinions.
- Compare charts between students or groups, discuss or debate.
- Have students watch TV commercials—bias is very prevalent in commercials. Evaluate the commercial's position—what bias is there? Why is Tide the best detergent?
- Propose a retort from a competing product—this requires students to look at both sides of a product and evaluate.

Chapter Three

Concluding the Project

Questions to consider

- What have you learned about bias so far?
- What is your opinion about the issue?
- Do you have facts to support your opinion?

Activities

- Rewrite the original article without any bias. Identify step by step:
  - What happens to the number of facts and details in the story?
  - What happens to the length of the story?
  - How easy or difficult is this task?
- Rewrite the original article changing the slant of the bias. Identify step by step:
  - What happens to the number of facts and details in the story?
  - What happens to the length of the story?
  - How easy or difficult is this task?
- Propose an outline, plan, or questionnaire journalists can use to check their news reports for bias.
- Journal overall feelings about bias in the media, news, commercials, TV etc.—how do students feel about it now?
- Hold a class discussion or debate once projects are over.

Resources

Books

The True Story of the 3 Little Pigs by Jon Scieszka; Puffin (March 1996)
The Three Little Pigs by James Marshall; Platt & Munk (October 2000)

Internet Sites

http://faculty.washington.edu/~jalbano/bias.html—This great site gives direct examples for students to identify bias in the news.

http://rhetorica.net/bias.htm—This site provides background on the many different types of bias found in the news.

http://www.mediaresearch.org/—This site believes the media are favorable to liberal views.

http://www.fair.org/—This site believes the media are favorable to conservative views.

In Search of Bias

From: Daily Messenger
Thursday
April 21, 2002

Previous Article:

Special City expects great new beginnings from XYZ Manufacturing Company. The company plans to relocate their 160,000 square foot rubber widget maker plant to Special City sometime next month. The company expects to bring 500 new jobs to the area.

New Article:

Citizens of Special City are concerned about the new 160,000 square foot manufacturing plant being built west of Special City. The rubber widget plant is expected to cause increased traffic on the Turnaround Turnpike. Citizens say the Turnpike is already unsafe.

Sample Re-write of an article.

Chapter Three

Language Arts
Grades 4-6

Newsy News

Newspapers are valuable resources for news from around the world. The specific format used for this delivery of information has common elements regardless of the country of origin for the newspaper. By understanding the structure and use of newspapers, students are able to translate that knowledge to develop their own school newspaper.

In the beginning stage of the project, students study the various types of news outlets, the structure of newspapers and the content within newspapers. Each activity helps build the foundation. By understanding the specific structure of a newspaper and identifying key elements, students will later use this information as a guide for their own newspaper.

During the developing stage, students will investigate and note the specific elements within a newspaper. While developing an understanding of each aspect, students will gain insight into the purpose and follow each aspect until the end of a newspaper.

The concluding stage will complete their learning by applying their knowledge to create a newspaper for their classroom or school. Specific literary elements may be applied depending upon curriculum needs and goals. By allowing students input into the news from their lives and organizing this information into a useable format, students learn to identify with themselves and their community.

Project Focus

Information is all mixed up in your classroom. Create a newspaper to set the story straight.

Why this project is valuable

Because we are living in the information age, students need to understand the sources of news available. Newspapers are one of the main avenues to learn information. This project is great for working with the whole class; each student will have a specific piece to complete for the paper. Maintaining the paper will enable you to continue teaching new concepts.

Students will learn

There are many possibilities. Tailor the project to fit your needs.

- Responsibility for their actions
- How to solve a problem
- Developing a useable plan
- Development of newspapers
- Structure of newspapers
- Purpose of newspapers
- Structure of news stories
- Story leads
- Use of dialogue
- Advertisements
- Other news outlets
- Parts of speech

Teacher's Tip

To maintain the paper, place it in the hands of a student(s) who enjoy the process and wish to continue publication.

Vocabulary

- Publisher
- Editor
- Deadline
- Lead
- Advertisement
- Headline

Teacher's Tip

Collaborate with a teacher in the younger grades using the Comic Relief project. The two projects could be used to create a full newspaper with comic strips.

What kids should know before beginning this project

Students should have some knowledge of newspapers in general.

Beginning the Project

Brainstorm and discuss

- Structure of newspapers.
- Different types of newspapers.
- Common elements of newspapers.

Questions to consider

- Do you have a local paper?
- What is the name and area the paper covers?
- Is there more than one area the paper covers?
- If so, how is the focus of each paper different?
- Who runs the paper?

Chapter Three

Activities

- Create project journal.
- Investigate other news outlets. Make a list of all you find.
- Investigate other newspapers and determine common sections. Make a list of all you find.
- Create an information scavenger hunt with the newspaper; include use of the newspaper index. Trade with a friend and perform the hunt.
- Rewrite two headlines after reading a story. Give each a different angle.
- Using a current news article, break it apart into who, what, when, where and why.
- Explain the sections and page numbering of the newspaper in your journal.
- Develop criteria to judge whether an article is well written. Display the criteria.

Developing the Project

Questions to consider

- Do students use newspaper at other times or only for this project?
- If used at other times, what for?
- What are other sources of news? TV, internet, the grapevine.
- Are students techno-savvy or have access to specialized computer programs?
- If so, could any be used to publish a newspaper?
- How do you decide if an article is well-written?

Activities

- Compare a daily print newspaper with a daily online news source. What are the differences and similarities? Make a chart.
- Research and detail on paper, poster or other method the job of the publisher, editor, reporter, pressman etc.
- Categorize the different types of articles in a newspaper such as:
  Headlines news, local news, personal interest, classifieds—find as many categories as possible.
- Explain in your journal why accepting news "through the grapevine" isn't the best method.
- Write five math problems based on information found in the newspaper.
- Research a current event found through another source.
- Research computer programs used to design newspapers such as: Adobe Photoshop, Illustrator, Pagemaker, and Quark Xpress.
  Explain what makes these such good programs for newspapers.

Concluding the Project

Questions to consider

- How often will the paper be published?
- Will it continue after this project?
- How many pages will it be?
- Is it color or black and white?
- Any photos, drawings, advertisements to be included?
- Have students worked together for one paper or have there been several groups creating different newspapers?

Activities

- Publish a full newspaper for your classroom or school.
- Provide comparison of two newspapers covering the same area. Document in a report, poster, multimedia presentation, etc.
- Write a detailed "To Do List" for someone wanting to start a newspaper.
- Using comic strips from younger grades, develop criteria to use in determining which strips are used in the paper and which are not.

Resources

- Invite a local newspaper publisher, editor, reporter or anyone related to the business to speak with the class.
- Visit a local newspaper for a field trip.

Books

The Young Journalist's Book: How to Write and Produce Your Own Newspaper by Nancy Bentley and Donna Guthrie; Millbrook Pr Trade (April 2000)

Kids in Print: Publishing a School Newspaper by Mark Levin; Good Apple (December 2000)

School Newspaper Adviser's Survival Guide by Patricia Osborn; Jossey-Bass (1998)

I highly recommend this book if you intend on developing a full newspaper for your classroom or school. Many great tips and ideas are included.

Internet Sites

http://www.onlinenewspapers.com/—This site allows you to search for papers all over the world.

http://www.ecola.com/—Another great source for multiple newspapers.

http://newslink.org/—More great links to newspapers. Nicely divided sections.

Newsy News

Volume: 1 Issue: 1 Page: 1

The Daily Tribune

In this issue:

- *School News*
- *Kim's Poetry Corner*
- *Dance News— Get the 4-1-1 here!*
- *Great Study Tips from Nick*
- *Demetri's Sports Stats*

Spring Dance Next Week!

From the Teacher's Desk:

√ *Don't forget about the vocabulary quiz on Thursday.*

Sample front page of Newspaper

Chapter Four

General Projects

Grades 2-3

Bicycle Extravaganza

Although no prior knowledge of bicycles is necessary for this project, most students have used bicycles in some capacity. By applying prior experiences or by building knowledge from the bottom up, students are able to learn a variety of concepts related to bicycles. These concepts may vary from basic knowledge to understanding the science of how bicycles work and the marketing behind the sale of them. Then by recognizing the differences, similarities and choices of bicycles, students form judgments as to their personal likes and dislikes. They form judgments and naturally the students learn about themselves. Honoring personal choices teaches students to feel confident and self-assured.

During the beginning stage, students are building a deeper understanding of the bicycles and factors related to them. By learning about bicycle features, where they are sold, and how sports associate with bicycles, students build their analyzing skills.

In the developing stage, students are required to place bicycles in the context of their real world. By investigating neighborhoods, safety elements, and warranty features, children relate bicycles to other aspects of life.

While in the concluding stage, students learn marketing and consumer awareness. These aspects coupled with using real data helps students to further understand bicycles and their world. Depending upon the ability level of your students, activities may vary to expand their understanding.

Chapter Four

Project Focus

Your grandpa is buying you a bike for your birthday. You want to make sure you get the best bike on the market. Create a market analysis for the five bikes he could buy you.

Why this project is valuable

As future consumers learning market and evaluation skills is essential. Consumer awareness is smart thinking.

Students will learn

- How to solve a problem
- Developing a useable plan
- Evaluation skills
- Bicycle safety
- Market analysis
- Warranty guarantees
- Bicycle features

Vocabulary

- Safety
- Helmet
- Unicycle
- Bicycle
- Markets
- Analysis
- Consumer

What kids should know before beginning this project

Students should have some knowledge of

- Safety features on bikes—reflectors, flags, bells, etc.
- Safety features for bikes—helmets, pads.
- Competition

Beginning the Project

Brainstorm and discuss

- How many kids own a bike?
- What kind of bike do they have?
- How old were they when they got their first bike?
- How did they learn to ride a bike?
- Who taught them?

Questions to consider

- What aspects of bicycles are the students going to study? Such as size, style, price, safety features, warranty.
- Do students know of various ways to buy a bike?
- Have students ever considered buying a bike online? Why or why not?
- What kind of bike do you like?
- Are there certain features you would like in a bike?

Activities

- Create project journal.
- Make a list of all the places you could go on your bike.
- Find five interesting facts about bicycles. Draw a picture to accompany the fact.
- Develop a list of various places to shop for bikes. Create a challenge for the most places.
- Explain in your journal the features you find most desirable on a bike.
- Design the "perfect" bike for you with all the features you could possibly want.
- Hold an award contest for the "Best Designed Bike."
- Investigate various sports that involve bicycles. Pick your favorite and tell why. For example: Bicycle Polo, Cycle Racing, Triathlons.
- Make a list of prefixes for bicycles such as uni-, bi- and tri-. Explain the meaning of each.
- Create graphs for various bicycle facts such as:
  - Number of kids who do/don't own a bike.
  - Age when kids first got a bike.
  - Type of bicycle kids have.

Developing the Project

Questions to consider

- What features of a bike will be studied?
- Is there a price limit?
- What features is Grandpa looking for in a bike for you?
- How long do you expect to keep this bike?

Activities

- Draw a map of your neighborhood designating where you would ride your bike.
- Make a list of common bicycle features.
- Obtain pictures of old bikes—compare them to new bikes. What's different/same?
- Create a "Bicycle Safety" poster.
- Write and perform a skit about "Bicycle Safety."
- Create a crossword or word search using bicycle words.
- Write five trivia questions about bicycle safety—switch with a partner.
- Design your own safety gear to wear on your bike. Color and label all pictures.
- Demonstrate with a chart or table the best warranty you find available on the bikes.

Concluding the Project

Questions to consider

- How many markets are available for students to investigate?
- What are the top features to be included in the analysis?
- Is there a price limit?
- How will students present their analysis?

Activities

- Create a chart in any format (poster, slideshow presentation, spreadsheet) for your grandpa showing the five different bikes and features. Include: size, style, price, safety features, warranty guarantee.
- Design and launch a bike safety plan for your school.
- Write an advertisement for your choice as the "Best Bike" convincing others it is the best.
- Design an order form for you to fill out and give to your Grandpa requesting the bike.
- Write the "Ideal Warranty." What would it include?

> **Teacher's Tip**
>
> For students' needing to be challenged—have them study the invention of the bicycle or the science behind how it works.

Resources

Books

The Big Bike Race by Lucy Jane Bledsoe; Holiday House (October 1995)

The Bear's Bicycle by Emilie Warren McLeod; Little Brown and Co. Reprint (November 1986)

The Evolution of the Bicycle by N. Wood; L-W Promotions (June 1996)

Internet Sites

http://www.nhtsa.dot.gov/kids/biketour/—This kid-friendly, easy to read, interactive site offers tons of tips on bicycle safety while students take a bicycle tour.

http://www.cdc.gov/ncipc/bike/kids.htm—The National Bike Safety Network for kids provides great safety tips and advice.

http://www.pedalinghistory.com/PHbikbio.htm—This site provides a quick history of bicycles. Early readers may need assistance in using this site or inputting the URL address.

Bicycle Extravaganza

Austin's Bicycle Mania

Name	Size	Price	Style
Mongoose Mountain Bike	Boys 14"	$129.00	Short Seat Mountain Tires
Quasar Howler	Boys 14"	$114.00	Full Suspension, Durable Frame
Jeep Commando	Boys 14"	$139.00	Good For Beginning Learners
Mongoose Rockadile	Boys 14"	$199.98	Can Ride On Dirt Or Pavement
Huffy Marengo	Boys 14"	$109.99	Front And Rear Brake

A sample market analysis.

General
Grades 2-3

Fire Safety

This study of fire safety can range from basic fire safety guidelines and prevention to understanding how to develop a fire prevention plan for your family. Students will take an active role in their own family or a fictitious family's home to prevent fires. Fire safety is essential for all students and general public alike. Learning the factors that lead to creating fires helps students to practice fire prevention.

In the beginning stage of this project, students develop an understanding of fire hazards and fire safety. By comprehending the dangers associated with fire and the necessary tools to use in the event of a fire, each child builds a foundation upon which they will later make decisions for their own fire safety.

During the developing stage, each student begins to recognize these dangers and safety needs by relating them to their personal lives. Each activity is meant to help personalize the information for the students. By involving family members, students are able to gain perspective of their own home.

The concluding stage offers students to take the information and use it to develop their own personal fire safety plan. Fire awareness helps students to develop safe habits and routines. It also gives them tools and ideas to use in the event a fire does occur.

Project Focus

You know fire safety is essential in a home. Establish a written fire safety plan for your home, including diagrams of how to exit your home. Share the plan with your family.

Why this project is valuable

Being able to focus and not panic in an emergency situation is a valuable skill for young students. Having a plan established helps to alleviate panicked feelings. Parents are usually grateful for this type of project.

Chapter Four

Students will learn

- Developing a useable plan
- Fire safety
- Proper response procedures
- Causes of fire
- Diagrams of buildings
- Using floor plans

Vocabulary

- Fire
- Safety
- Extinguish
- Help
- Assist
- Prevention
- Hazards
- Alarms

What kids should know before beginning this project

Students should have some knowledge of

- Dangers of fire.

Beginning the Project

Brainstorm and discuss

- What causes a fire?
- What should you do in case there is a fire?
- Who do you call when there is a fire?

Questions to consider

- Do you have a school fire plan?
- Is a map posted somewhere in your room detailing the exit manner for fire drills?
- Have you had a fire drill this year?
- Does your family have a fire plan?

Activities

- Create project journal.
- Make a list of possible reasons for a fire.
- Write a checklist of fire safety tools for your home.
- Conduct a fire safety systems check of your home.
  - Do you have fire extinguishers?
  - Do you have fire alarms and /or smoke detectors?
  - Does your family have a plan?
- Find five fire facts you didn't previously know.
- Draw a picture with the facts to demonstrate the facts.
- Poll parents on fire facts. Create a "Did you know" dialogue. Report what parents know.

Developing the Project

Questions to consider

- Are parents cooperative with this project?
- Do students understand need for subject?
- Are there any special considerations for students such as bars on windows, two-story homes, etc.

Activities

- Design attractive signs to be posted in your home showing emergency exit avenues.
- Write a survey asking parents about their fire safety tips.
- Compare tips from several parents to create a master list.
- Create a fire slogan or chant for your family. Present to all family members.
- Draw pictures and label possible fire starters in your home.
- Practice writing home address and phone number several times. Use fun colors and writing techniques. (Check with parents for permission—do not post private information.)

Concluding the Project

Questions to consider

- Do students have a floor plan or diagram of their home?
- Have students had the opportunity to work with parents on this project?
- Are students growing in fear or alleviating fear? Work to alleviate.

Chapter Four

Activities

- Create a list of steps to follow to report a fire. Include who to call, what to say, vital home information, and any other relevant information.
- Make a flow chart depicting the steps to take to get out of a burning house.
- Draw a map of your home using a floor plan or help from parents. Use a red arrow to show possible escape routes for each member of the family.
- Write a step by step list, report, poster etc. of things to do/not do when responding to a fire. Include Stop-Drop-&-Roll, neighbors' house to run to, common meeting place outside, etc.
- Role play—"Responding to a Fire in Your Home"—for your classmates, include a written script.

Resources

Contact your local fire department. Invite a fireman to talk to the class and give prevention lessons and activities for kids to follow in the event of a fire.

Books

Stop, Drop and Roll (A Book about Fire Safety) by Margery Cuyler; Simon & Schuster (September 2001)

No Dragons for Tea by Jean Pendziwol; Kids Can Press (March 1999)

"Fire! Fire!" Said Mrs. McGuire by Bill Martin, Jr.; Harcourt Brace (1996)

Internet Sites

http://www.nfpa.org/sparky/—Follow the Family Stuff link in this site to obtain a grid to use when developing a floor plan a diagram for an escape route. Lots of other great stuff in here.

http://www.dos.state.ny.us/kidsroom/firesafe/firesafe.html—This site is written for kids and contains rules and guidelines for fire safety.

http://www.kidsafetyhouse.com/—This site provide background and information relative to fire safety in the home. It is written for kids.

Teacher's Tip

If floor plans are not available for students, obtain floor plans from the Internet or a floor plans book. Allow students to choose one to use in creating an escape route diagram.

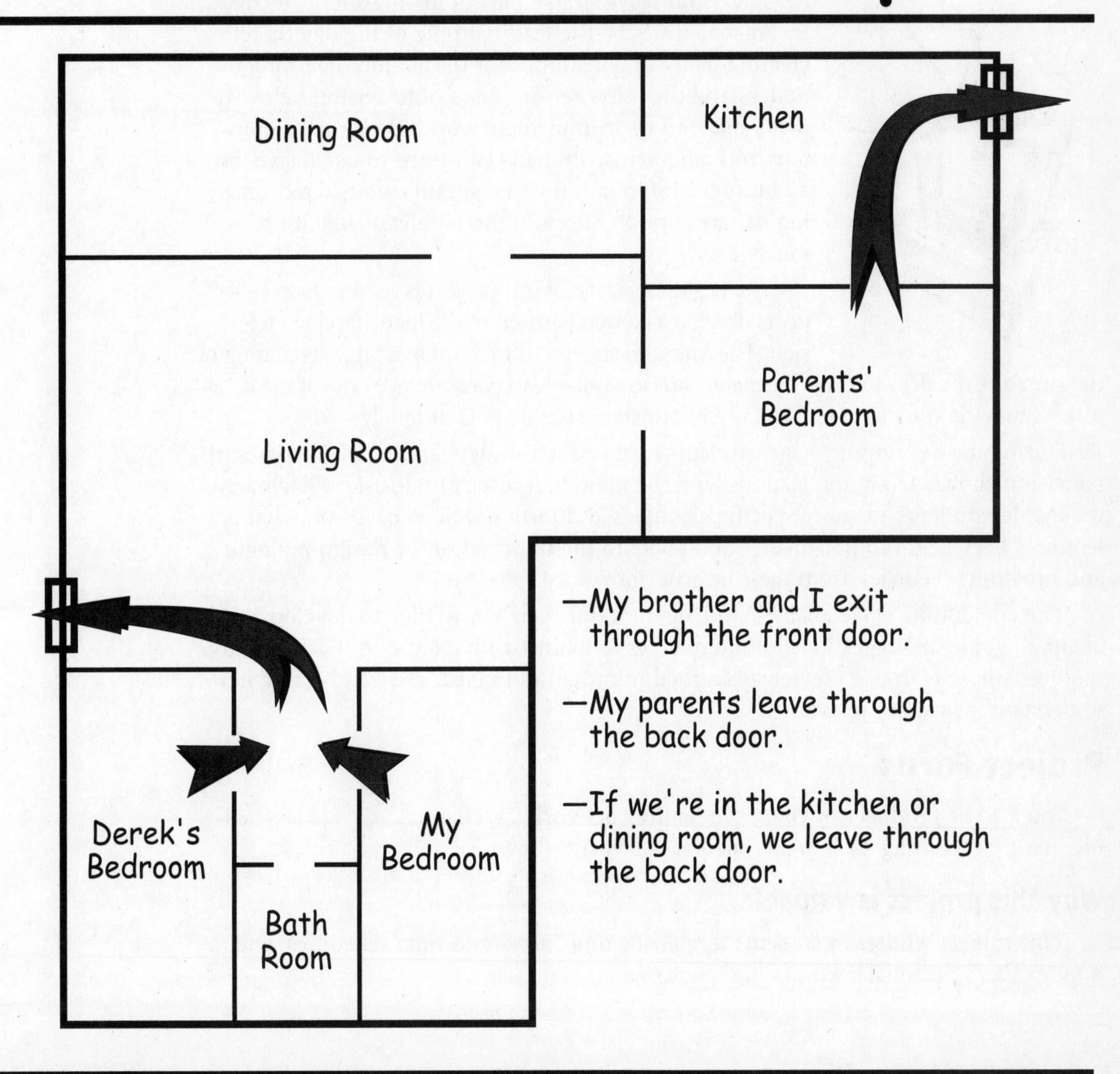

A sample fire plan.

Chapter Four

General
Grades 3-5

Family Broadcasting Network

Television plays a large part of many Americans lives. It can sometimes be the leading topic of the day thereby creating many opportunities for discussion. Learning to understand the purposes and uses of television helps students relate to their immediate world. By forming opinions and judgments, students can learn to use television as a tool for information while simultaneously recognizing patterns and trends, and the pitfalls of watching too much TV.

The beginning stage of this project is used to help students develop a comprehension of the many facets of television. The activities are geared to develop an understanding of the structure of television and the many choices available to consumers. Some of the activities require the student to poll and use information specific to their family.

During the developing stage, students further their understanding of the types of television shows, television stations, and the many purposes of television. Each activity enables students to study specific examples and form opinions based on what is learned. They begin to personalize and relate to the information by designing logos and providing feedback from their favorite show.

The concluding stage enables students to input their knowledge to develop a TV channel, guide or suggestions for alternatives to mainstream television. Each activity requires students to use previously learned information to customize and create new understandings about television.

Project Focus

You are the producer of your own family's network TV channel. Create a schedule line-up including each family member's favorite shows.

Why this project is valuable

This project will teach students to manage time, adhere to time schedules, and honor other's personal opinions and choices.

Students will learn

- Time management
- History of television
- Structure of television programs
- Types of programming
- Scheduling tactics
- Emotional attachment to television

Vocabulary

- Television
- Production
- Commercial
- Schedule
- Program

What kids should know before beginning this project

Students should have some knowledge of

- The structure of televisions programs
- Various time allotments
- Various subjects
- Various channels focusing on specific subjects

Beginning the Project

Brainstorm and discuss

- What are your favorite TV shows?
- Why are they your favorite?
- What is one show you don't like that someone else watches?
- Why don't you like that show?
- How much TV do you watch each day or week?

Chapter Four

Questions to consider

- How many members are in your family?
- How long will each person have for their choice of programs?
- What are drawbacks or benefits of television programs?
- Do you prefer fiction or nonfiction television programs?

Activities

- Create project journal.
- Give three reasons television programs are good and three reasons why they are bad?
- Write an advertisement for your favorite program, convincing others to watch it.
- Explain the premise of your favorite television program, including basic plot line, characters, timeline, and other relevant and appropriate aspects of the program.
- Write a survey asking your family about their favorite TV programs.
- Conduct the survey and graph or table the results.
- Make a list of all the types of television programs, i.e. talk shows, comedy, cartoons, cooking shows, news, sports events, concerts, music video shows, history, etc.

Developing the Project

Questions to consider

- What is the name of your TV channel?
- Do you know any TV related people to talk to or bring to class?
- How does TV make you feel?
- What does TV offer your family?

Activities

- Design a logo for your channel.
- Explain the logo in your journal.
- Write a description of each of the types of shows. Give a real example of each.
- Tell the location, what the letters stand for and popular shows on each of the major networks.
- Write a proposal for a new show that should be on TV.
    - What would it be about?
    - Why would others watch it?
    - What age group would it appeal to?
- Compare two sitcoms, cooking shows, sports events, etc. Tell how they are alike and different.
- After watching an "educational" show, explain what could be learned from watching that show.
- Watch your favorite show and time how long each commercial break is. Then calculate how long your show really is.

Concluding the Project

Questions to consider

- How much overall TV time will be included for the week?
- Are the students choosing reasonable time amounts?
- Have students considered each member of the family?
- Have students considered other factors such as bathing, homework, dinner time in their schedules?
- What is the overall purpose of your network?
- What is special about your network?

Chapter Four

Activities

- Create a complete TV guide for your family, noting time schedules and each individual's favorite choice in programs, include a description of each program.
- Propose to others in house, alternative programs to watch based on like/dislikes found in the surveys.
- Write a proposal for each member of the family stating what each could do with their time rather than watch TV
- Create a comparison chart of satellite and cable television.
    - What is the cost of each?
    - What equipment is needed for each?
    - How many channels do each offer?
    - What is the best choice for your family?

Resources

Books

Hi There, Boys and Girls: America's Local Children's TV Programs by Tim Hollis; University Press of Mississippi (November 2001)

TV Mania: A Timeline of Television by Judith Henry; Harry N. Abrams (October, 1998)

The Great TV Turn-off by Beverly Lewis; Bethany House (January 1998)

Rudolph the Red-Nosed Reindeer: The Making of the Rankin/Bass Holiday Classic by Rick Goldschmidt; Miser Bros Press (October 2001)

Internet Sites

http://www.tvshow.com/tv/shows—This site provides information on all the types of programs and some information about each. Great for building background knowledge.

http://www.tvguide.com/—This is the official TV Guide site. You can put in your zip code and get specific channel programming.

http://dmoz.org/Arts/Television/Programs/—Comprehensive site that links to just about anything about television you want to know, including information on the major networks.

Family Broadcasting Network

Lauren's TV Guide

Person	Time	Favorite Program	Notes
Dad	Sun 12:00-6:00pm	Football	Dad watches all Sunday
Mom	Tue 8:00-9:00pm	'24'	Mom never misses this
Sister	Every day 4:00-4:30pm	'Sister, Sister'	My sister and I fight about this. We could alternate days
Me	Every day 4:00-4:30pm	'The Cosby Show'	
Brother	Every morning 8:00-8:30am	'Power Rangers'	Everyone is gone from the house, except Mom

A sample family TV Guide & Description.

Chapter Four

General
Grades 3-5

Personal Holiday

Each country, family, heritage, and religion has many holidays specific to it. The meaning behind each of these significant events helps students to develop an understanding for their world and where they fit in it. By understanding significance of events, students are able to relate to significant events in their lives.

The beginning stage of this project is used to build an understanding of holidays, the history behind them, their purpose, and specific details related to them. Each activity investigates specific aspects of holidays. By investigating these aspects, students develop an understanding of the reasons for and rituals behind holidays.

During the developing stage, students relate these common elements to their own lives. Each activity helps students identify what is important to them. They personalize the key features of a holiday and designate similar ideas to create their own holiday. By creating their own holiday each child develops a voice for their own opinion.

In the concluding stage, students create a display of their favorite holiday. They develop reasoning and rituals behind their holiday. Each holiday is shared and explained. The child develops a sense of themselves, their world and significance within it through sharing holiday rituals.

Project Focus

All holidays have specific meaning tied to them. Often there are specific stories, symbols, or events that occur with each holiday. Design a holiday other than your birthday that is all about you or something you love. Detail the significance and celebratory need for this day.

Why this project is valuable

Understanding the significance of celebrated holidays helps students to understand their world. Designating a personal holiday instills importance and celebrates individuality in students.

Students will learn

- Significance of certain holidays
- Types of celebrations
- History behind holidays
- Self-importance
- Individuality
- Symbolism
- Symbols relative to holidays

Vocabulary

- Holiday
- Festival
- Festivities
- Celebration
- Culture
- Symbol
- Emblem

What kids should know before beginning this project

Students should have some knowledge of

- Holidays in general.
- Reasons for celebrating various holidays.

Beginning the Project

Brainstorm and discuss

- Create a list or web of all the holidays students know.
- Discuss importance of each or most.
- Give exact dates or months of each.
- Discuss emblems or symbols representative of holidays—i.e. Christmas Trees.

Chapter Four

Questions to consider

- Do you have students of varied ethnicity in your class?
- If so, what unique holidays do they celebrate?
- What is the significance of these holidays?
- Do students have a "favorite" holiday?
- If so, what is it?

Activities

- Create project journal.
- Create a list of reasons for holidays—religious event or person, to honor people, remember significant date or event, etc.
- Make a list of holidays to fit into each category of reasons—i.e. Easter and Christmas are religions holidays, etc.
- Research a specific holiday from another country. Give the specific details of the day.
- Explain in your journal your favorite holiday. Tell what makes it special.
- Create a graph of favorite holidays for your class. Record graph in journal.
- Write a letter to a friend explaining your favorite holiday.

Developing the Project

Questions to consider

- What is the significance of your holiday?
- Would other people be interested in this day?
- Do any particular objects or items easily associate with this holiday?
- Are there any specific colors that associate with the subject for this holiday?

Activities

- Design an emblem, symbol, or ornament for your holiday.
  Explain its significance.
- Make a list of symbols or items usually identified with common holidays. Draw pictures and explain each symbol.
- Choose a day on the calendar to be your specific holiday. Explain why you chose this day.
- Write a letter to a friend convincing them to celebrate your holiday.
- Design a poster or other visual display of your holiday.
- Write a story detailing how your holiday came to be.
- Draw a post card of your holiday. What specific details would be included on the postcard?

Concluding the Project

Questions to consider

- Have students chosen realistic subjects to celebrate?
- Do students have symbols or emblems to associate with their holiday?
- Can students detail the significance of each holiday?

Activities

- Design your own holiday. Detail the significance and key aspects to the holiday. Include date and any special symbols or emblems.
- Present to your class a display of your holiday and how you will celebrate it.
- Write a letter to the President of the United States, explaining why your holiday should be a national holiday.
- Design an ad campaign for your new holiday. How will others find out about it?
- Role-play with a friend the introduction to your new holiday.

Resources

Books

Kids Around the World Celebrate! The Best Feasts and Festivals from Many Lands by Lynda Jones; John Wiley (November 1999)

Children Just Like Me: Celebrations! by Anabel and Barnabas Kindersley; DK Publishing (October 1997)

Internet Sites

http://www.earthcalendar.net/—This site is a database that allows you to investigate holidays based on date, religious affiliation, or country. Lots of basic facts here.

http://holidayfestival.com/—This site lists holidays from around the world for each month of the year. Use for activities investigating holidays in other countries.

http://www.billpetro.com/HolidayHistory/—This site is written by a historian and provides the history behind American holidays.

http://www.essortment.com/in/Holidays.History/—Another great site to find out history, details, legends, and facts about holidays. This site can easily inspire many activities.

> **Teacher's Tip**
>
> Religious holidays can develop tricky discussions. Agree to honor each student's and family's beliefs ahead of time.

Personal Holiday

National Chocolate Day!

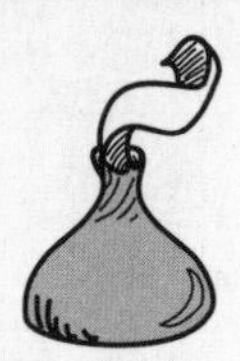

February 15, 2003

Use your Valentine's Day candy!

Menu

Breakfast
Scrambled Chocolate
—All your favorite chocolate bars crumbled in a bowl! Yum!

Lunch
Chocolate Soup
—Melted hot chocolate with cream.

Dinner
Chocolate Casserole
—Chocolate with graham crackers, nuts and marshmallows

Activities	What to do
1. Chocolate Design	Mold your favorite character out of chocolate.
2. Chocolate Stash	Play hide & seek with your favorite chocolate bar.
3. Chocolate Melt	Predict which piece of chocolate will melt first in your hand. Get your prediction right and win a PRIZE!!!

A sample holiday design.

General
Grades 4-6

Exercise Mania

Discussion about exercise and types of exercise is prevalent in every magazine, newspaper, and television program. Sometimes conflicting information is confusing. By investigating exercise and the many factors related to it, students will develop an informed opinion that they can relate to their own needs, likes, and dislikes.

During the beginning stage of this project, students learn about the many forms of exercise including indoor and outdoor activities. They will also investigate their family's typical physical routines, and the general time schedule used for the exercise. Students learn to investigate other possible times for working-out after learning their family's regular routine. As a final step in this stage, each child educates themselves about the health benefits of exercise.

The developing stage of this project allows students to categorize and explain the benefits of exercise. They begin to identify with exercise and the benefits by fully investigating various types of exercises. Students will also learn about their family's personal likes and dislikes where exercise is involved.

In the concluding stage of this project, the class is required to take all previous knowledge—bring it together to make judgments and decisions based on their family's preference for exercise. If a student's family is uncooperative, these activities may need to be for a fictitious family. They'll learn to use time schedules to reach goals for themselves and their family members who will in turn—learn the value of a regular exercise routine.

Project Focus

You keep hearing about the health benefits of exercise. You know your family needs to get more exercise into their daily routines. Design a family exercise plan that all members can enjoy.

Why this project is valuable

The issue of physical fitness can be confusing due to the many conflicting reports available. Teaching students to use the information to develop a plan that works for them will lead them to develop their own healthy lifestyle.

Chapter Four

Students will learn

- Purposed of physical fitness
- Benefits of physical fitness
- Types of exercises
- Sports as part of physical fitness
- Developing time schedules
- Honoring family's wishes

Vocabulary

- Exercise
- Sports
- Aerobic
- Non-aerobic
- Health
- Physical Fitness
- Interval

What kids should know before beginning this project

Students should have some knowledge of

- Types of formal exercise.
- Various sports for exercise.
- Family's general weekly schedule—consult parents if necessary.

Beginning the Project

Brainstorm and discuss

- Favorite activities for kids to play.
- Exercises that feel like play.
- Types of exercise students dislike.
- Any health or physical disabilities students or family members might possess.

Teacher's Tip

If students intend on really taking this plan home for family members, please remind them any person with health or physical issues should seek medical advice.

Questions to consider

- Do students regularly play sports?
- Do parents regularly play sports?
- Do students exercise currently?
- Do parents exercise currently?
- Do students enjoy exercise?

Activities

- Create project journal.
- Draw a web showing exercises for inside and outside.
- Write a survey for parents asking regular exercise routine if any.
- Obtain doctor recommended exercise amounts. Compare to current routine from parents.
- Write a weekly time schedule for family. Show all current activities and obligations.
- Suggest possible exercise times based on current time schedule.
- Conduct research and report on five ailments caused by lack of exercise.
- Create categories for exercise. Include cardiovascular, weight training, yoga, and stretching. Give examples for each category.

Developing the Project

Questions to consider

- Do any family members need special considerations?
- Do any family members need special equipment?

Activities

- Find five health benefits to exercise. Explain each in a journal.
- Explain in a report, poster, multimedia presentation etc. the difference between aerobic and non-aerobic exercise and the benefits of both.
- Write a survey asking family members what exercises they are willing and not willing to do.
- Compile survey into possible/not possible choices.
- Write a diary entry from the viewpoint of someone who exercises regularly. Explain how good you feel.
- Make a chart depicting number of calories burned for thirty minutes of various exercises. Use the last website listed for information.
- Use chart to form judgments about what is the best exercise for their family. Record in journal. Explain your choices.
- Write a letter to a friend detailing the benefits of exercise.

Chapter Four

Concluding the Project

Questions to consider

- Are students experiencing help or resistance from home?
- Will students really use this or is it for a fictional family?
- Are students an "expert" in certain exercises or sports?

Activities

- Design a time table and exercise plan for each member of your family. Include days, times, and type of exercise for each member.
- Design a fitness journal for each member of the family. Allow room for entries and personal notes.
- Propose five possible ways members of your family can incorporate more exercise into their daily routine. Write each on a 3 x 5 card and give to person.
- Hold a trial exercise day where each child or group teaches classmates how to perform certain exercises or play a specific sport.

Resources

Invite a personal trainer, doctor, or other health professional to talk to the class about the benefits of exercise.

Books

Fit Kids....Getting Kids "Hooked" on Fitness Fun! by Mandy Laderer; Allure Pub (February 1994)

Babar's Yoga for Elephants by Laurent de Brunhoff; Abrams Books for Young Readers (September 2002)

Five Kids & A Monkey Solve the Great Cupcake Caper: A Learning Adventure About Nutrition and Exercise by Nina M. Riccio; Creative Attic (July 1997)

Internet Sites

http://www.kidshealth.org/kid/—Lots of basic information including games and activities written for kids.

http://exchange.co-nect.net/Teleprojects/project/Fitness—This fabulous site provides background and has teacher resources and activities for kids.

http://www.caloriecontrol.org/exercalc.html—Students can use this site to investigate the calories burned for each type of activity. This site is great for creating charts and comparing information. Give students a quick run through. It also provides many activities students might not consider to be exercise.

Exercise Mania

Stephanie's Fitness Journal

Day: ______________________________

Date: ______________________________

Type of exercise	
How long	
How I felt	
Personal notes and goals	

A sample Journal Design.

Chapter Four

General
Grades 4-6

Mall Delight

What teenager doesn't dream about hanging out in a mall? This project gives students the opportunity to learn about one of their favorite spots while learning valuable budgeting skills and time management. It also introduces students to the business side of something they often consider only for entertainment.

In the beginning stage of this project, students begin to investigate malls in a different format than just walking or shopping in a mall. They'll contact local malls to obtain behind the scenes information to use throughout the project. Each child needs to investigate malls as a business with many factors and differences for each type of mall.

For the developing stage of this project, students personalize the project by researching a friend's personal likes and dislikes. Each friend will receive a gift from the type of stores where they usually shop. Students will use this information to determine when and where each present will be bought. All friends and gifts are fictitious so each child can adjust as needed.

During the concluding stage, students develop a full plan to accomplish the guidelines. They'll learn time management, adhering to others' likes and dislikes, and creating a useable plan with a set purpose.

Project Focus

You have two hours to buy ten friends a gift. No friends like gifts from the same store. Design a plan to complete your task in the time frame needed.

Why this project is valuable

Learning to develop plans and use resources readily available is an essential skill for all students. Paying attention to feelings, wishes and wants of other people develops compassion for all participants involved.

Students will learn

- Structure and design of buildings
- Floor plans
- Time management
- Developing a time line
- Marketing strategies

Vocabulary

- Outlet
- Upscale
- Online
- Marketing
- Purchase

What kids should know before beginning this project

Students should have some knowledge of

- Malls in general.
- Purpose of mall.
- Other shopping entities.

Beginning the Project

Brainstorm and discuss

- Students previous visits to malls.
- Favorite and least favorite parts of malls.
- Parents thoughts on kids attending malls.

Questions to consider

- Have students been to a mall before?
- Do they like to go to the mall? (This is almost a silly question for teenagers, but by asking it you will get them thinking about what they like and don't like about it.)
- What types of malls are there? (Indoor, outdoor, outlet, upscale, strip malls, online)

Activities

- Create a project journal.
- Investigate all the malls in your city, area, region. Make a list of each one and type.
- Write to or call the mall asking for an overview map of the mall showing location and names of stores.
- Compare two or three types of malls designating similarities and differences. Research number of stores, total square footage, types of stores, clientele.
- Make a list of each mall (and their location) in your city and note how many times you go there to shop.

- Write a poll asking peers, parents etc. about favorite type of mall.
- Write a defensive paper explaining which mall is the best to shop.
- Write a journal entry explaining your favorite shopping experience.

Developing the Project

Questions to consider

- What do your friends like?
- What don't your friends like?
- How much do you want to spend on each friend?
- Could stores in the mall be categorized?
- How would you categorize them?

Activities

- Make a list of the ten presents you will buy.
- Explain why you chose what to buy each friend.
- Use a chart to show which friend gets what present and the store you will shop.
- Categorize the presents. What category does each present fit in?
- Categorize the stores in the mall. How can you do that?
- Pretend you won a shopping spree at your local mall. Detail how much you won and what you will buy.
- Make a list of sale signs and other marketing strategies used at the mall.
- Make a list of five things you don't like about the mall.
- Research the history of malls. Report your findings.

Concluding the Project

Questions to consider

- Have students been to the mall lately?
- If not, how long has it been?
- Is the mall's business office available for questions?

Activities

- Document the entire plan including:
  - A diagram of the mall with stores labeled.
  - Which stores you will shop.
  - What type of store it is.
  - Which order of stores you will shop.
  - How long you will spend in each shop.

- Write a time-line itinerary for your shopping experience.
- Design a webcast for Mall of America's online site.
- Write a letter to a friend explaining why you were unable to get them the gift the wanted.
- Design your "dream" mall. What would it include/not include? What would it look like?

Resources

Any local mall or shopping center.

Internet Sites

http://www.mallofamerica.com—Mall of America's official website. There is tons of information on this site that could be used to generate questions or ideas for the project. Each idea could be used to make the project more specific to your curriculum needs.

http://yellowpages.msn.com/simplesearch.aspx?KWD=shopping+malls—This site will allow you to input your city or town and pull up shopping malls within the area.

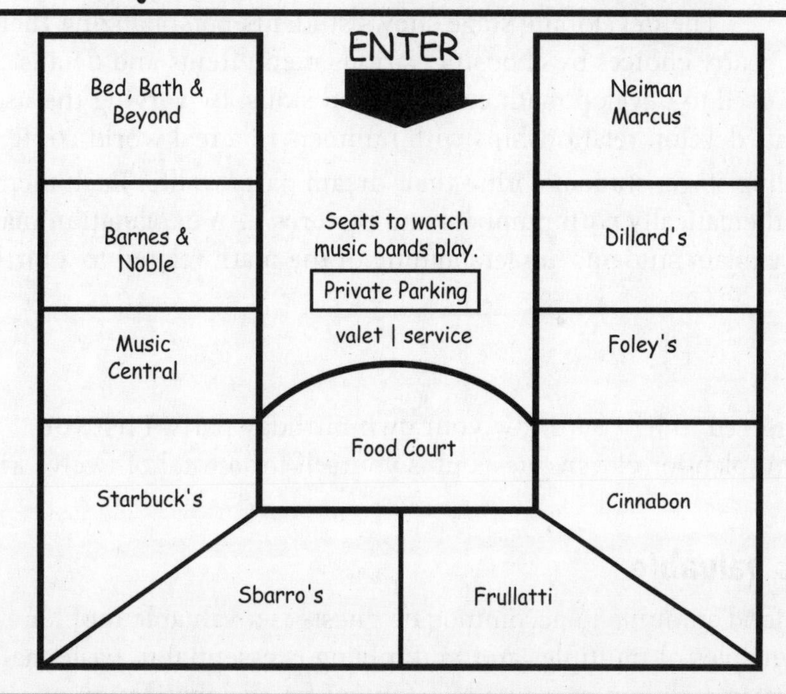

A sample "dream mall".

Chapter Five

Math Projects

Grades 2-3

Birthday Party

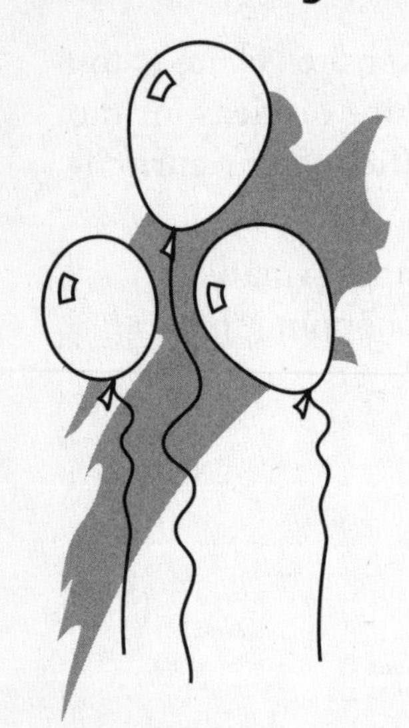

Kids love birthday parties. They mostly love their own, but revel in the joy of others' parties too. Some students have a simple party every year while others never have an elaborate party. Regardless of their prior experiences, students love to discuss and learn about such an exciting subject. This fun subject can be used to learn many interesting math and numeral concepts.

At the beginning stage, students learn about the various types of birthday parties. They are allowed to dream and investigate various options available. Basic numeral concepts are introduced to develop an understanding of how math is a part of any party planning.

The developing stage shows students personalizing their party choices by choosing various menu items and details. Each food item is then used to develop math concepts and skills. By varying the use of the numbers, students develop relationships with numbers in a real world context.

During the concluding stage, students bring their dream party to life. Each menu item is represented mathematically with numbers and pictures. An explanation may be provided to further explain students' understanding of the math related to a birthday party.

Project Focus

Your parents have agreed to let you throw your own birthday party. First you have to design the menu plan for eleven guests plus yourself for a total of twelve kids at the party.

Why this project is valuable

Learning to adjust food amounts to accommodate guests is a valuable tool for students. Using the principles of multiples and multiplying is essential to basic math understanding. This project will give students real context for the application of these skills.

Students will learn

- Responsibility for their actions
- How to solve a problem
- Developing a useable plan
- Real use of multiples
- Multiplication as repeated addition
- How to develop a menu
- Categories of food
- Relevance of numbers with food
- Serving sizes
- Reading food labels

Vocabulary

- Serving size
- Menu
- Food label
- Multiplication
- Multiples
- Repeated addition

What kids should know before beginning this project

Students should have some knowledge of

- Basic numeral concepts:
- Adding
- Subtracting
- Multiples—if this has not been discussed previously, it can be approached at the appropriate times during the project

Teacher's Tip

At the end of the project, plan one big birthday party with foods kids can cook in class. Make it a complete learning activity for all involved. Make sure to recruit some reliable parents to help.

Chapter Five

Beginning the Project

Brainstorm and discuss

- Previous birthday parties' students have had.
- Food at the parties.
- Number of guests at the parties.
- Any themed parties?
- Any problems at the parties?
- Favorite part of the party.

Questions to consider

- What would be your idea of a fantastic birthday party?
- How many guests would you invite to your party?
- Would you have a theme at your party?
- What kind of food would you have at your party?
- Would the food go along with the theme?
- What if one of your guests didn't like your choice of food?
- What would you feed them?
- Who will buy the food?
- Who will make the food?

Activities

- Create project journal.
- Create a web showing many different types or themes of birthday parties.
- Chose a theme or idea for your party and defend why you chose it in your journal.
- Draw an advertisement for your party.
- Compile a list of possible foods to be included at your party.
- Write number sentences for the number 12—such as 3+3+3+3 =12; write as many as possible.
- Eliminate all number sentences that don't fit multiples format.
- Write math word problems that equal 12.

Developing the Project

Questions to consider

- What foods have the students chosen?
- Are these foods easy to work with?
- Are these foods easy to multiply?
- What special requirements might the students have for their food?
- Any food allergies?
- How are the foods packaged?
- How many packages of food will be needed to feed 12 kids?

Activities

- Compile a list of all food expected to be served at the party. Record the list with pictures and descriptions in journal.
- Compare serving sizes for two products—i.e. hot dogs and macaroni and cheese on paper, poster board etc.—use a t-chart.
- From the comparison, estimate how many servings of the food would be needed. Draw a picture representing the food and number of servings.
- Write math problems for each food product—i.e. I have 12 guests. There are 10 hot dogs and 8 buns per package. How many packages of each will I need?
- Write a paragraph explaining how to make one of the foods included at the party.
- Create a time sheet detailing the events of the party including when each food will be served.

Concluding the Project

Questions to consider

- Have the students come across any real obstacles in planning?
- Were the foods chosen realistic?
- What information will students need to complete their menus?
- How developed are your students' planning skills?

Chapter Five

Activities

- Write a detailed menu for your guests showing what will be served.
- Create a display of the food showing how many servings of each food will be fixed and served.
- Draw a diagram of where each person will sit at the table. Make sure to include all 11 guests and yourself.
- Change the diagram to look completely different—draw the new sitting places. Encourage students to use groups of 3/4/6 etc.
- Write a step-by-step plan for the food preparation to give your parents including all relevant math information such as quantities and serving sizes.

Resources

Books

Peas and Honey: Recipes for Kids (With a Pinch of Poetry) by Kimberly Colen; Boyds Mills Pr. (January 1995)—This is a fantastic book with tons of information.

The Best Birthday Parties Ever! A Kid's Do-It-Yourself Guide by Kathy Ross; BT Bound (August 2001)

Pretend Soup and Other Real Recipes: A Cookbook for Preschoolers and Up by Mollie Katzan and Ann Henderson; Tricycle Press (May 1994)

Internet Sites

http://www.eezplanit.com/planning.htm—This site is great for students to see what all goes into planning a birthday party.

http://www.healthychoices.org/act-kit/toc.html—Recommended for teacher use. This site offers a variety of activities for kids. Check under Part 1, Activity 2 for a downloadable guide to meal planning.

http://family.go.com/parties/birthday/specialfeature/partycentral_intro_ parties_sf/—What a long address! You may need to help little one's type all this, or use this site for ideas to pass along. Lots of great party tips here!

Birthday Party

Jessica's Birthday Party!

At my party, I want to serve hot dogs.
I have 11 guests and me. That equals 12.
I want to give all of us 2 hot dogs each.

2 + 2 + 2 + 2 + 2 ☐ ☐ ☐ ☐ ☐

12 guests
x 2 hot dogs
24 hot dogs total

I will need at least 24 hot dog buns also!

8 buns per package
x 3 packages
24 Total buns

A sample problem presented graphically.

Chapter Five

Math
Grades 2-3

Mighty Measures

Measuring items can be difficult to understand for many students. Using different measuring methods and tools can further complicate the issue. Understanding various methods of measurement helps students to grasp the need for standardization when measuring. It also helps students understand which measurement units to use for various items.

In the first stage, students develop an understanding for two specific measuring systems—standard and metric. This understanding will include the various units within each system and common items measured with each unit. Students study the common words and prefixes that denote specific numeral amounts.

During the development stage, students will gain a greater understanding of the need for standardization in measurement. They will use and label measurement units for common items found in their world. Each opportunity affords students the chance to investigate the relationship of these items to the unit of measure.

For the last stage, they must transfer their knowledge of measurement systems to propose a new measuring system together. This system will be unique to each student or group. Justification must be given as to the reason for choosing the measuring device. Through this process students will begin to notice the need for consistent measuring devices and modes.

Project Focus

You just don't understand all those meter and inches. Create a new standard for measuring.

Students will learn

- Responsibility for their actions
- How to solve a problem
- Developing a useable plan
- Metric measurement
- Standard measurement
- Standardization in measurement
- Realistic measuring tools

Teacher's Tip

Students might struggle with what to use as a measuring device. Provide examples such as noodles, feathers etc.

Vocabulary

- Metric system
- Standard or universal system
- Measure
- Measurement

Specific measurement names such as

- Meter
- Inch
- Foot
- The websites listed below have many vocabulary opportunities.

What kids should know before beginning this project

Students should have some knowledge of

- Measurement in a broad sense.
- Difference between liquid and dry measure.

Beginning the Project

Brainstorm and discuss

- Why we measure.
- Things we measure.
- What might go wrong if we didn't measure?
- Measuring names students are familiar with already.
- Compare 1 oz. dry and 1 oz. liquid measure.

Questions to consider

- Do students understand the difference between liquid and weight measure?
- What objects are available for measure?
- Do students cook? Relate measures to cooking if possible.
- What are uses for each standard of measurement? i.e.—when do we use feet, miles etc.

Chapter Five

Activities

- Create project journal.
- Research and report on the history of the metric and standard systems of measurement.
- Create a chart showing full word and common abbreviations for each system.
- Using both metric and standard, measure and label common objects in classroom.
- Using a blank map, label which measuring system used is in various countries.
- Weigh various items. Write statements such as, "The shoe weighs more than the pen."
- Draw or obtain pictures of various objects. Label beside object which unit of measure would most likely be used when measuring that object.

Developing the Project

Questions to consider

- Do students understand there are two major systems of measurement used in the US?
- Do students understand how each developed?
- Do students have any experience using the metric system?
- Are any students from other countries that use the metric system?

Activities

- Write a journal entry explaining which measuring system you prefer—standard or metric. Explain which seems easier and why.
- Read aloud *How Big is a Foot* by Rolf Myller.
- Create a chart depicting what went wrong with measuring system in book.
- Write a journal entry explaining why we need standards in measuring.
- Create a Venn diagram comparing inches and meters. In the middle, the commonalities of the use for measurements.
- Create conversion chart from inches to meters. Include actual measurements and picture representing each measurement.
- Make a list of possible new measuring devices.
- Write a vocabulary test using measuring words and trade with a partner.

Concluding the Project

Questions to consider

- What are the writing abilities of your students?
- What technology is available?
- Could pictures be cut from magazines instead of drawn?
- Do students understand standardized measurement?
- What does that mean to their system?

Activities

- Provide a detailed explanation of your measuring system, include pictures.
- Create a conversion chart depicting your measuring system to either the metric or standard measurement.
- Make a diagram of your classroom using measurements from your system.
- Write a letter to your parents, teacher, newspaper etc. explaining why your system is a better system for measuring.
- Provide a chart showing various systems of measurement and changes to measurement over time.

Resources

Books

How Big is a Foot? by Rolf Myller; Random House Childrens Pub.; Reissued (August 1991)

The Fattest, Tallest, Biggest Snowman Ever (Hello Math Reader—Level 3) by Michael Rex et al; Cartwheel Book (February 1997)

Tell Me How Much It Weighs (Whiz Kids) by Shirley Willis; Franklin Watts, Inc. (December 1999)

Internet Sites

http://lamar.colostate.edu/~hillger/—The US Metric Association official site provides unlimited background and statistical information including free resources for teachers.

http://www.unc.edu/~rowlett/units/—This fantastic site provides links to definitions of prefixes and definitions of measurement words.

http://www.convert-me.com/en/—This site allows you to enter specific values that can be easily converted to other systems—by filling in one value, many others are provided.

Mighty Measures

Donna's Measuring System

I propose we use chocolate for our measuring system.

Just think... How many M&M's would it take to weigh my Mom's car?

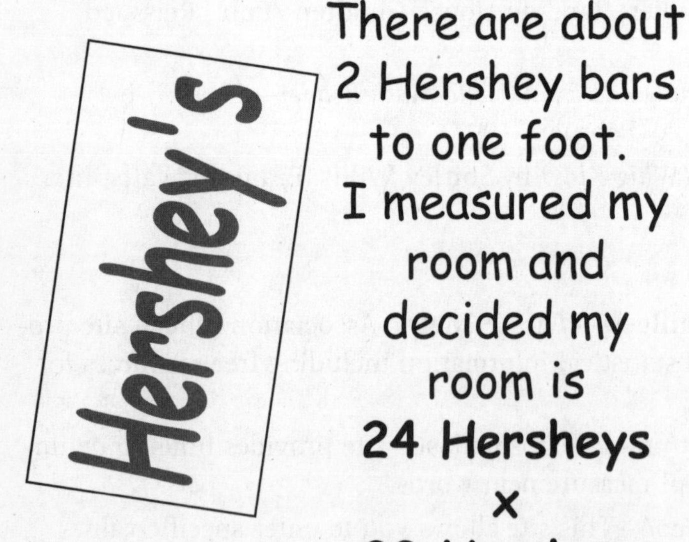

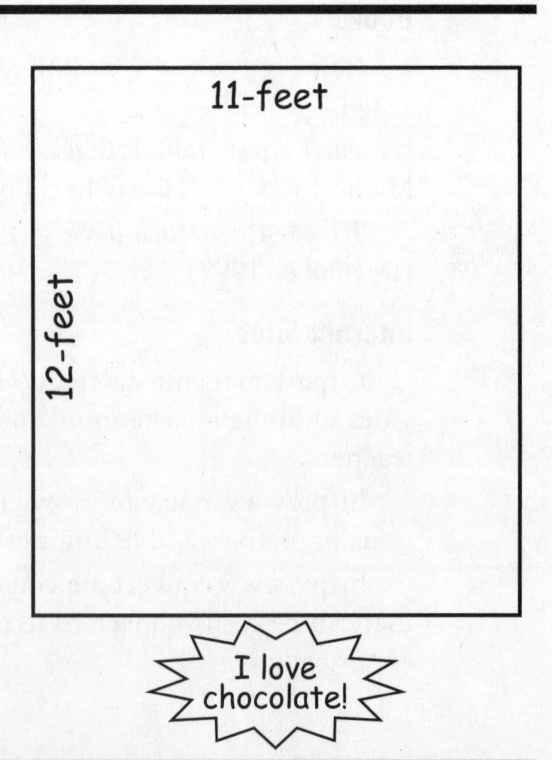

A sample proposal.

Math
Grades 3-5

Allowance Budget

Money is an exciting subject for most students. Students rarely have a good understanding of the various denominations and uses for money. Helping students understand money, its denominations and uses, aids students in developing good money sense. Helping students understand work in relation to earning money develops positive work ethics in children.

In the beginning stage of this project, students learn to develop an understanding for the various denominations of money and the possible avenues to earn money. Students need to determine types of chores and time alloted to complete each chore. These factors will be used to establish an allowance proposal.

For the developing stage, students are required to personalize the information by gaining input from other students and their parents. They will learn valuable graphing and survey skills while collecting this data. Then they must begin to form opinions based on their findings. These opinions will guide their overall proposal.

During the concluding stage, students need to develop a full allowance proposal. This proposal will include their choice of chores and money to be earned. Students may include suggestions for uses of their money and potential savings plans.

Project Focus

Your parents have decided to give you an allowance. Justify what you will do and how much you will earn for your allowance. Provide your parents with a proposal. All chores and money amounts must be reasonable.

Why this project is valuable

Students will be responsible for their own money. Understanding money issues in relation to work is a necessity. Children need to develop an understanding of how to earn money and use it wisely. This is a great chance to instill a positive work ethic.

Chapter Five

Students will learn

- Responsibility for their actions
- How to solve a problem
- Developing a useable plan
- Money denominations
- Budget guidelines
- Fiscal responsibility
- Work ethics
- Principles of money management
- Time schedules
- Time management
- Money management

Vocabulary

- Allowance
- Money
- Currency—if you choose to study monetary measures from other countries
- Chores—Duties—Jobs—what term will you use?

What kids should know before beginning this project

Students should have some knowledge of

- Days of the week
- Money

Beginning the Project

Brainstorm and discuss

- Money—denomination of each coin and bill.
- Other currencies.
- Personal weekly schedules—each home differs.

Questions to consider

- What is an allowance?
- What can be done with an allowance? Buy things and services, save, invest.
- What are chores?
- What can be done for chores?
- Do the students have any regular chores now?
- How much money is a lot, how much is a little?
- How or who will decide this?
- How busy is your family during the week?

Activities

- Create project journal.
- Create a chore web—what are some examples of chores?
- Make a chart showing coins and bills and how much each is worth?
- Write a weekly time schedule based on what students know—or send a note home asking parents to fill out weekly time schedule and have students rewrite it.
- Use the time schedule to decide time of day chores could be completed.
- Make a list of chores students are interested in completing.
- Write a brief description of each.
- Draw a picture of each activity.

Developing the Project

Questions to consider

- What is a reasonable allowance appropriate for their age?
- Are there any restrictions to what you can do?
- Who will pay the allowance?

Chapter Five

Activities

- Create a survey—ask the class what a reasonable allowance is.
- Create a survey—ask parents what they feel a reasonable allowance is.
- Compare surveys—draw t-chart or Venn diagram.
- Translate survey to a bar or line graph.
- Write a paragraph defending chosen amount of allowance.
- Repeat steps for chores –
    - Survey students.
    - Survey parents.
    - Compare surveys.
    - Translate to graphs.
    - Write paragraph defending choice of chores to complete.
- Journal findings.

Concluding the Project

Questions to consider

- How much time is the student expecting to spend on chores?
- How much money is the student expecting to earn from allowance?
- What is the final decision on the chores the students intends on completing?
- What will the student do with the money earned?

Activities

- Write a paragraph, letter, note, or a proposal of some sort—explaining chores and earnings from student to parent.
- Create a poster showing proposed chores and earnings.
- Video tape a verbal proposal while demonstrating chores.
- Make a chart showing how much money will be saved, spent, and invested.
- Chart savings over a one month and then a twelve month period.
- Compile a list of possible big purchases or projects to be done with the saved money.

Resources

Books

The Case of the Shrunken Allowance by Joanne Rocklin; Cartwheel Books 1999

Earning an Allowance by Joy Berry; Gold Star Publishing Inc.

The Kids Allowance Book by Amy Nathan; Walker & Co. 1998

Internet Sites

http://www.kidsbank.com/index_2.asp—This site provides good background information on money and banking.

http://www.kidsturncentral.com/links/moneylinks.htm—This site provides link to many other valuable sites to teach kids about money and what to do with it.

http://www.chartjungle.com/—This site offers charts to download for free, yes some you can buy, but many are free. These charts can help students develop a plan for their chores. There are also time chart to use to teach about time. Lots of great FREE stuff here.

Teacher's Tip

This would be a fantastic opportunity to study currency from other countries or previously used barter trade economies. Kids get a kick out of trading donkeys for food etc. Use your imagination and guidelines of this project to think where you can go with a further currency study. Create charts comparing the dollar, yen, euro, etc.

Allowance Budget

Derek's Allowance Budget

★ **Clean Room** 1 x a week	=	**$2.00**
★ **Take Out Trash** 4 x a week	=	**$2.00**
★ **Empty Dishwasher** 4 x a week	=	**$1.00**
Total		**$5.00**

BONUS:	Help Dad rake leaves (it's a big job!)	=	**$5.00**
	Help Mom with laundry (I hate laundry!)	=	**$3.00**

A sample proposal.

Math
Grades 3-5

Math Games 101

Games are a valuable resource for fun and learning. Most students gain an understanding of strategies and concept development used in games. This enables students to understand real math concepts used in their world. By using current games as a model, students are able to translate new math understandings into a game format. Developing and building games also releases a student's creativity.

In the beginning stage, students must find a way to discover the appropriate directions when playing games. They will also uncover the many types of games and the variables specific to the different games. They need to begin to look at math concepts as part of games and how these can be incorporated into a game. Students will also achieve an understanding of their classmates' personal likes and dislikes.

During the developing stage, students personalize what they have learned by designating specific skills and objectives to their game. Each aspect of their game will be considered individually. Once the individual pieces are considered for the game, then they will be able to bring all the pieces together to create a completed game.

For the last stage, students will present their fully conceived game. Each game may be modeled after another game or be completely new. Students will use math concepts as part of their game. These concepts may vary depending upon the ability level and needs and goals of the curriculum.

Project Focus

Design a math game that requires using multiplication, division, probability or any other grade level math concept. Include instructions on how to play the game and any necessary game pieces.

Why this project is valuable

Designing games not only gives students the opportunity to work with math in a new light, it raises creativity. Each game should have an overall skill or concept it teaches or uses.

Chapter Five

Students will learn

- Responsibility for their actions
- How to solve a problem
- Developing a useable plan
- Purpose of games
- Rules of games
- Strategic planning
- Computation
- Developing rules
- Various math algorithms
- Math vocabulary

Vocabulary

- Addition
- Subtraction
- Multiplication
- Division
- Strategy
- Computation
- Any other math vocabulary you spotlight for this project.

What kids should know before beginning this project

Students should have some knowledge of

- Playing games
- Rules of games
- Purpose of games—fun and competition.

Beginning the Project

Brainstorm and discuss

- Types of games—board, computer, card, electronic, video, field.
- Equipment for various games.
- Strategy or rules behind various games.

Questions to consider

- What types of games are there?
- What is a student's favorite game?
- What do they like about games?
- What do they wish they could change?
- Which games require the least/most equipment?
- What math concept do you want the students to learn?

Activities

- Create project journal.
- Try to play a game without giving any directions. Discuss difficulties with this approach. Discuss what would help if directions are not given.
- Create a word search, crossword puzzle, anagram or any other word game using math vocabulary.
- Obtain instructions for several board games. Compare the length of instructions, equipment needed, number of players, and estimated time for the game as well as any other particular aspect that would fit your curriculum.
- Write a journal entry explaining your favorite game. Why is it your favorite?
- Write a survey asking classmates what their favorite games are.
- Conduct survey and graph result.
- Write five math problems based on your survey.
- Write five math problems based on games. What games will you use?

Developing the Project

Questions to consider

- What is the basis of the game?
- What is the strategy?
- What is needed to make the game?
- Any special game pieces or board pieces?

Activities

- Establish guidelines for your game. State objective, rules, number of players, and way to win.
- Make a list of vocabulary for your game.
- Write a vocabulary quiz for your game.
- Write a letter to a friend explaining in your own words how to play a math game you find on the internet.
- Describe and design the pieces and parts of your game, explaining the use of each.
- Write a game review, explaining the educational aspects of your game and why others should "buy" it.
- Create a list of math objectives one must know to play your math game.
- Explain, using words and numbers, how to solve problems like the ones in your game.

Concluding the Project

Questions to consider

- What format will students use to present the games?
- Are the games feasible?
- What materials are needed?
- Any special considerations for the game?

Activities

- Make a full version of your game—play it with your friends.
- Create an official game guide, explaining the game, how many players, how to win, necessary equipment and any clue or tips needed to be a winner.
- Hold an official game day where each student presents his game and classmates rotate through playing the games.
- Create a catalog your games. Color and include prices and brief descriptions, including educational opportunities from each game.
- Make a "cheat sheet" of quick operation tools to use when playing games. What clues will you include to help other students?

Resources

Games

- Any game with instructions for students to explore.

Books

Math Curse by Jon Scieszka; Viking Childrens Books (October 1995)

Math for Kids & Other People Too! By Wide World Pub Tetra (November 1997)

Anno's Math Games—Book I, II and III by Mitsumasa Anno; Philomel Books (1987)

Internet Sites

http://www.funbrain.com/—This site has tons of great educational items for kids. Follow the numbers link to play various math games.

http://www.coolmath4kids.com/—This fantastic site contains tons of valuable and fun information for kids about math including many games. Use this site for the letter activity in the developing stage.

http://www.toonuniversity.com/default.asp—Another great site with games to play and use as a guide in developing other games.

Math Games 101

Drew's Money Mania

Objective: To be the person at the end of the game with the most money! Cha-ching, Baby!

Number of Players: 2-5

Supplies Needed: Dice & Gameboard

Rules:
1. Roll dice.
2. Move forward that number of spaces.
3. Add each money amount you land on.
4. Whoever has the most money at the end wins.

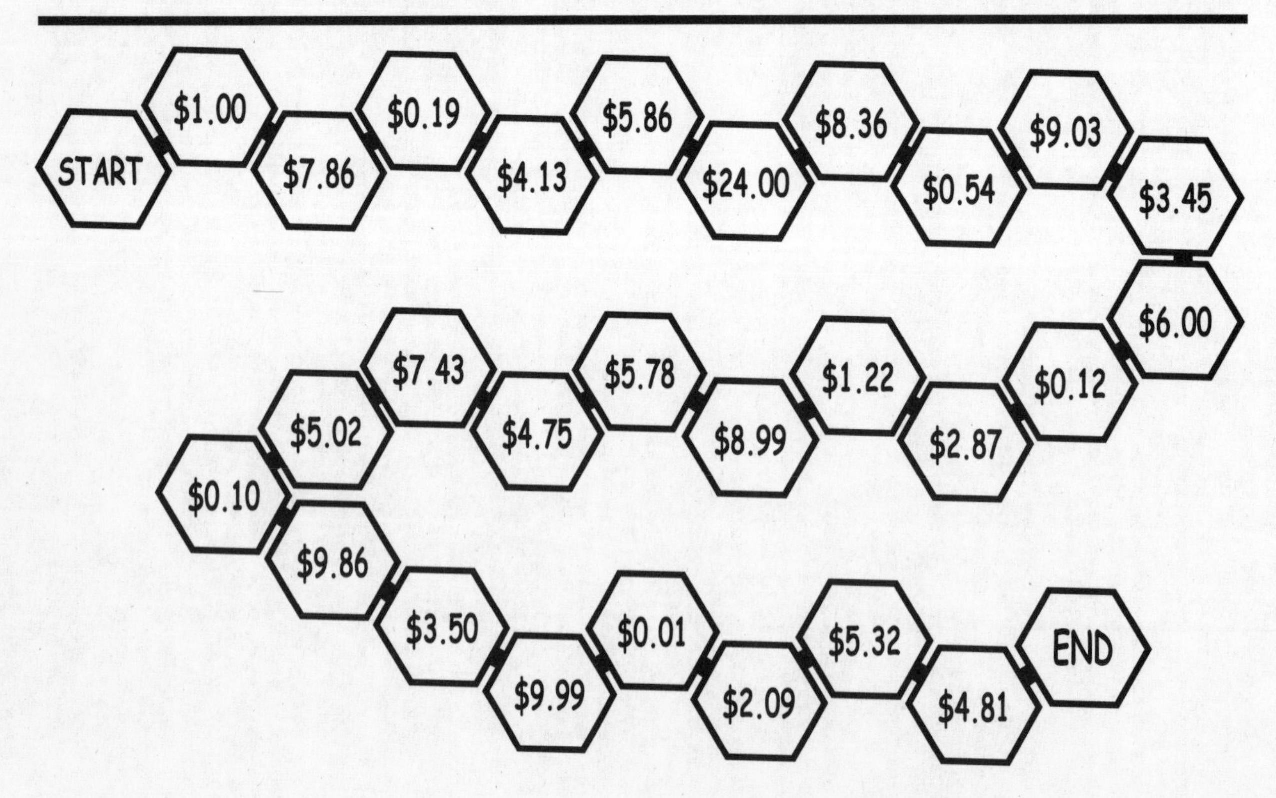

A sample game and gameboard.

Math
Grades 4-6

Dad's Diet

Teaching students to develop healthy habits early in life helps them to learn how to maintain a well-rounded lifestyle later in life. It helps students to understand how to maintain a wholesome diet if they understand all aspects of food and nutrition. By studying common maladies caused by food deficiencies, students develop an understanding for proper and beneficial nutrition.

In the first stage, students gain knowledge of basic food and nutritional information. They'll use current government data and guidelines as a resource. By tying nutritional information with specific foods, students can create a visual idea of a healthy diet. By understanding food deficiencies, students can work to develop food choices to combat these ailments.

During the second stage, each child customizes their learning to meet the specific needs of a real or fictitious person. Each person represents a nutritional deficiency or other diet related illness. By studying various ailments, students learn disease causing factors and possible preventive measures. This will support their learning about the benefits and inclusion of specific foods within any diet regime.

In the third stage, students use their knowledge to propose possible new diet choices for the ill person to follow. The culmination may result in a sample one-day diet or various other suggestions to follow for diet changes.

Project Focus

Your dad has been put on a diet. Keeping with the dietary guidelines set by his doctor, create a sample one-day menu. This project will require students to make basic scenario decisions in order to proceed to the next step.

Why this project is valuable

Students are continually bombarded with a wealth of information. Conflicting news reports can leave consumers confused about what types of food and how much of a food should be eaten within one day. Students need a broad understanding of the dietary guidelines to help them make informed decisions about their own health and eating habits.

Chapter Five

Students will learn

- Decision making
- Responsibility for their actions
- How to solve a problem
- Developing a useable plan
- Application of multiplication
- Application of division
- Percentages
- Dietary guidelines
- How to read charts and graphs

Vocabulary

- Carbohydrates—use in body
- Proteins—use in body
- Fat—use in body
- Fiber—use in body
- FDA—Food and Drug Administration
- Nutrition

What kids should know before beginning this project

Students need to have some knowledge of

- Basic food groups.
- Basic multiplication and division concepts.
- Calculator skills.
- How to find percentages—even if only on a calculator.

Beginning the Project

Brainstorm and discuss

- What is nutrition?
- What are the basic food groups?
- Why is it important to know about food groups?
- How can students use this information in their lives?

Questions to consider

- Why has the doctor put your dad on a diet?
  - Can be made-up—such as, too much fat in diet, not enough fiber, high cholesterol, heart disease.
- What does your dad eat now?
  - Make up or poll real dad or adult male with poor eating habits.
  - Students must be reasonable with scenarios.
- What are the dietary restrictions set by his doctor?
  - To make it simple use the standard guidelines for a male from the Food and Drug Administration.

Activities

- Create journal for project.
- Read aloud book(s) about food groups, exercise, FDA guidelines.
- Journal thoughts about topic so far—include students' opinions.
- Create web or graphic organizer of information, include in journal.
- Create an outline of current foods eaten by dad.
- Compare current foods eaten to FDA guidelines with T-Chart.
- Categorize foods by what currently fits these guidelines and what doesn't.

Developing the Project

Questions to consider

- What is the goal of the diet?
- Based on goal, what foods does this include or not include?
- Are these changes life-long or temporary?
- What is needed in the new diet?

Activities

- Contact a local doctor to come speak to the class on best ways to manage certain aspects of diet.
- Research specific health issues related to new dietary needs—write brief report on disease or illness.
- Write a letter to an organization requesting further information on the disease or illness.
- Collect a variety of food labels.
  - Practice reading them and breaking down information into percentages of carbohydrates, fat, protein, and calories. This can be done by taking total grams of fat, protein, carbs, or fiber and dividing it by total grams of food.
- Use dietary guidelines and food information sources to determine dad's typical diet configurations now.
  - Break down each food by fat, carbohydrates, fiber, protein and calories. Give total percentages for each category, including total calories consumed each day.
- Using FDA guidelines, create a multimedia presentation of what foods are best and worst for dad to eat.

Concluding the Project

Questions to consider

- Is dad willing to make drastic changes to his diet or only small changes?
- Depending on dad's willingness, what type of plan will you propose, complete change or slight variations?
- Are all foods available in your area?
- Are some foods seasonal?
- What are some substitutions for foods if original is not available?

Activities

- Create a one day example of exact foods to eat at each meal including snacks.
- Create a comprehensive list of foods to choose from for each meal including snacks.
- Create a comparison chart showing previous one day diet and new one day diet.
- Formulate charts projecting health benefits of new diet.
- Draw pictures and explain possible diseases contributed to by an unhealthy lifestyle.

Resources

- Contact your local doctor for information on dietary needs.

Books

The Complete Book of Food Counts by Corinne T. Netzer; Dell Pub Co; 2000

Food Rules! The Stuff You Munch, Its Crunch, Its Punch, and Why You Sometimes Lose Your Lunch by Bill Haduch; Puffin, 2001

The Edible Pyramid by Loreen Leedy; Scott Foresman; 1996

Internet

http://www.cfsan.fda.gov/~dms/educate.html—This is the official Food and Drug Administration site for kids, teens, and educators. There is lots of fun activities and information for students here.

http://www.healthfinder.gov/—This site provides opportunities for students to look up information as needed. There are many research possibilities here. It is a good place to begin.

http://www.usda.gov/news/usdakids/—This site provides comprehensive information on the food guide pyramid and breakdown of foods.

Dad's Diet

Diagnosis: Dad has acid reflux disease.

★ TYPICAL DIET Before Changes	★ TYPICAL DIET After Changes
Breakfast Eggs w/picante sauce Bacon Orange juice	**Breakfast** Eggs **no** picante sauce Wheat toast Apple juice
Lunch Chili cheese burger w/ onion French Fries Soda	**Lunch** Chicken sandwich Pretzels Water
Dinner Pizza —loaded	**Dinner** Steak Salad Baked potatoes —no extras

Also: Drink less soda.
Drink more water.

A sample proposal of diet changes.

Math
Grades 4-6

Statistic Savvy

Statistics are everywhere. Labels, billboards, television, radio, and newspapers all regularly relay information using statistics. Often these statistics can be misleading if you do not completely understand what information is being represented and what that information means. Children are particularly susceptible to confusion because of their age and naiveté.

The beginning stage of this project will establish the many uses of statistics in sports and music record sales. Students will learn the use of these statistics—along with the math and numbers behind each statistic. They'll also learn what the statistic means and why it is measured.

In the developing stage, students need to personalize their learning by choosing a specific sport or musical outlet to study. Then they demonstrate and explain the statistics specific to that avenue. Students will further relate statistics to their school and/or important music or sports heroes.

During the concluding stage, students demonstrate their understanding of statistics specific to their sport or music arena. This information may be represented through various formats chosen by you or the students. Then they have the opportunity to explain statistics in their own words to students in a younger grade. Students may also perform a scavenger hunt for statistics reporting as many as they can find.

Project Focus

Your head is swarming with facts and statistics. You see statistics everywhere, including when you play sports or listen to your favorite music. Create a display representing specific statistics. Explain what each statistic means to your sport or musician.

Why this project is valuable

As part of the information age, data is everywhere. Learning to decipher and use data and statistics is essential to building an understanding of our world.

hapter Five

Students will learn

- Responsibility for their actions
- How to solve a problem
- Developing a useable plan
- Data sources
- Statistics in everyday life
- Understanding statistics
- Using statistics
- Specific statistical information
- Statistics and sports
- Statistics and music

Vocabulary

- Statistics
- Data
- Trends
- Depending on the sport, various words could be used—such as for baseball:
  - Batting average
  - Runs batted in
- Depending on area of study in music

What kids should know before beginning this project

Students should have some knowledge of

- Calculating averages
- Multiplication
- Division
- Calculator skills
- Percentages

Beginning the Project

Brainstorm and discuss

- What are statistics?
- Where are statistics found?
- What sports use statistics regularly?
- What statistics are common to your sport or music choice?
- What are some ways statistics are calculated?
- What are some ways statistics are used?

Questions to consider

- Do any of the parents use statistics on a regular basis?
- How do they use statistics?
- Do students play any sports or follow specific genres of music?
- Do they know any sport statistics off the top of their head?
- Do students know who is No. 1 on the billboard charts this week?

Activities

- Create project journal.
- Watch a major sports event on TV. Listen for statistics. Write down at least two.
- Explain what the statistic you heard means in your journal.
- Make a list of all the sports that use statistics as a measure of an athlete's success.
- Locate statistics on your favorite baseball player. Explain what each means.
- Compare two baseball player's statistics. Make a t-chart.
- Explain who the better player is based on statistics.
- Create a vocabulary list and game to learn words unique to your sport.
- Research the numbers behind the Billboard's music rating scale. Explain in your journal.

Developing the Project

Questions to consider

- What is your favorite sport?
- Why is it our favorite?
- Is it a statistic driven sport?
- If not, should you choose one that is?
- What is the most statistic driven sport?
- How are statistics used to determine success of pop/rap/rock/country stars?

Activities

- Make a list of common statistics used in your sport.
- Give actual example of each.
- Explain the math behind the statistic.
- Write a letter to the top star in your sport.
- Interview a coach from your school. Obtain statistics about your school's sports.
- Create a display of the statistics from your school.
- Compare five "hot" artists. Explain the statistics used to determine their rank on the billboard charts.
- Write five math problems using the statistics behind the billboard charts.

Chapter Five

Concluding the Project

Questions to consider

- Did students choose the sport or music route?
- Did they find adequate information?
- Was the information useable?
- Do they understand the math behind the statistics of their choice?

Activities

- Create a full display of the top statistics used in your sport, explaining each of them.
- Compare two sports players or musicians, detailing the statistics behind their success.
- Explain why statistics are valuable and what you learned during this project.
- Write a letter to a student in a younger grade, explaining the use of statistics in your sport or for your musician.
- Create a graphic display of various statistical data related to your favorite sport or music genre. Label and explain why you choose those statistics.
- Using a newspaper, highlight all statistical data you can find.

Resources

Books

A Cartoon Guide to Statistics by Larry Gonick and Woollcott Smith; HarperCollins (February 1994) This book provides great background for teachers or high-end kids—not for the average or below-level student.

Time for Kids Almanac 2003 with Information Please by Holly Hartman; Time, Inc. Home Entertainment (July 2002) Joel Whitburn is an author that has compiled many books on billboard hits and statistics. Check with your local library for any books available by him.

Internet Sites

http://davidmlane.com/hyperstat/—This site provides great background. Follow links to find examples and explanations of basic statistical information. Lower achieving students may not be able to make sense of it all, but certainly more astute students will find this site valuable.

http://www.infoplease.com/sports.html—A virtual almanac of sports statistics.

http://www.billboard.com/billboard/index.jsp—This site has all the latest music charts and tons of data related to movies and music. From this site, students can choose what statistics to study and investigate further.

Statistic Savvy

Keith's Cowboys

Statistic Name	Who It's For	Statistic	What It Means
Passing Yards	QB Quarterback	300 Yds	When a QB passes over 300 yds, it is a great game!
Rushing Yards	RB Runningback	100 Yds	A RB that rushes more than 100 yds is great!
Fumbles	Any player w/ ball	Varies each game	A fumble is dropping the ball—it's not good.
Turnovers	Opposing team	Varies each game	Turnovers are when the other team gets the ball. Not Good!

A sample explanation of statistics.

Chapter Six

Science Projects

Grades 2-3

Invention Alley

Inventions make the world-go-round. Each day new items, ideas, and systems are introduced. Some are completely original ideas while others are improvements on existing items. Regardless of the reason for the invention each new invention creates change. Each student has the opportunity to create change through their invention. Students tend to be creative and full of ideas at this age. For those not as inquisitive, giving them direction and basic understandings regarding the process of inventions often releases their creativity.

While in the beginning phase, students will investigate the many inventions, gadgets and ideas of their immediate world. They'll create lists and criteria for these inventions by reading steps other inventors have taken. These steps aid the students in developing their own inventions during the concluding phase. By understanding how other students develop their inventions, they can model their thinking following the same pattern.

During the developing phase, each child learns about other kid's inventions. They will investigate the reason behind inventions and the various uses for inventions. Then the student writes-up patent information and how to make an invention a reality.

At the concluding stage, students will design their own invention. The invention may be presented in many formats including an actual prototype. It may also be a detailed account of the invention on paper. Students might choose to represent the invention with pictures or try to sell the idea to other classmates. Each outcome will be based on the needs and ability level of your students.

Project Focus

You've recently invented something to make your life easier or better. Describe your invention and present a prototype.

Why this project is valuable

Students that learn to interact with their world, gain understanding of their world. By investigating inventions and inventing themselves, students open their creative minds to the endless possibilities.

Students will learn

- Responsibility for their actions
- How to solve a problem
- Developing a useable plan
- Process of invention
- Previous inventions
- Previous inventors
- Creative thought
- Types of inventions

Vocabulary

- Inventor
- Invention
- Patent
- Trademark
- Prototype
- Product
- Service

What kids should know before beginning this project

Students should have some knowledge of

Some previous inventions that are part of our everyday life are:

- Telephone
- Indoor plumbing
- Electricity
- Roller blades

Chapter Six

Beginning the Project

Brainstorm and discuss

- What are some inventions you know?
- What might be your favorite invention?
- Why is it your favorite?
- What is something you wish worked better?
- What would make it better?

Questions to consider

- What is the socioeconomic status of your kids?
- What is the exposure level of your kids?
- What resources are available for kids to practice inventing?
- Do you know any inventors, engineers, project managers?
- What are the types of inventions? Product and service—discuss each.

Activities

- Create project journal.
- Using donations of old toys, tear apart a toy and put it back together as best you can.
- Explain in your journal how the toy works including drawings.
- Read aloud *Galimoto* by Karen Lynn Williams. Discuss thought process of book.
- Make a list of the steps Kondi took for his invention, including materials he collected.
- Watch TV commercials. Make a list of ten new inventions advertised. Include the name, type of invention, purpose of the invention.
- Categorize the above inventions into product or service category. Make a t-chart.
- Write a survey to ask your parents and friends about what they would like to see invented.
- Conduct the survey and graph results.

Developing the Project

Questions to consider

- What are common inventions students are aware of?
- Is background available on these inventions?
- What characteristics do inventors possess?
- What characteristics do your students possess?
- How much time is available to research other inventions?

Activities

- Make a list of ten inventions you benefit from. Categorize them as product or service.
- Detail who invented the above inventions, the type of invention and who else uses them.
- Using the survey from the beginning, propose new invention ideas.
- Compare two kid inventors. Make a chart showing age when invented, type of invention, help they received from parents.
- Make a list of the steps the kid inventor followed when creating their invention.
- Research patents. Determine if your invention needs a patent. Record in journal.

Concluding the Project

Questions to consider

- What resources do students need?
- Are these individual projects or group projects?
- What part is each student contributing?
- How involved is the student's invention?
- Is it realistic and feasible?
- Does it require any special materials?

Chapter Six

Activities

- Present your invention to the class including prototype.
- Present a full plan of your invention including:
    - The type of invention it is.
    - Why the invention is needed or useful.
    - Who will use the invention?
    - What is needed to make the invention?
    - A sample drawing of the invention.
- Write a letter to XYZ Company stating your idea and why they should use your idea.
- Perform a skit or role-play for the class a major inventor's life and his inventions.

Resources

Invite an inventor, patent attorney, engineer, project manager, or anyone else involved in the process of inventing or improving products to the classroom to discuss their job. They may inspire young inventors or give clues to the process etc. Design your criteria based on their job description.

Books

Galimoto by Karen Lynn Williams; (Mulberry 1990)

Gizmos and Gadgets: Creating Science Contraptions That Work (& Knowing Why) by Jill Frankel Hauser; Williamson Publishing (1999)

Brainstorm! The Stories of Twenty American Kid Inventors by Tom Tucker; Sunburst (October 1998)

The Kids Invention Book by Arlene Erlbach; Lerner Publications Company (May 1999)

Internet Sites

http://inventors.about.com/—This About.com page offers many links to invention statistics and information.

http://whyfiles.larc.nasa.gov/text/kids/Research_Rack/tools/idea_collection.html —This long address takes you to a NASA page with a document that can be printed for students to use when determining the next steps in their invention. Print this out for the serious inventors in your class.

http://kids.patentcafe.com/how-to-invent/index.asp—On this site, students can walk through the steps of invention online. The site walks students through step by step. This site is great for good readers, but slow readers may struggle with the necessary reading.

Invention Alley

Heidi's Amazing Drink Finder

November 20, 2002

X 1/2 Manufacturing
5000 Industrial Way
Anytown, USA 99999

To Whom It May Concern:

Sometimes I forget where I put my drink. I invented a device that I hook to my wrist. With this device, I can...

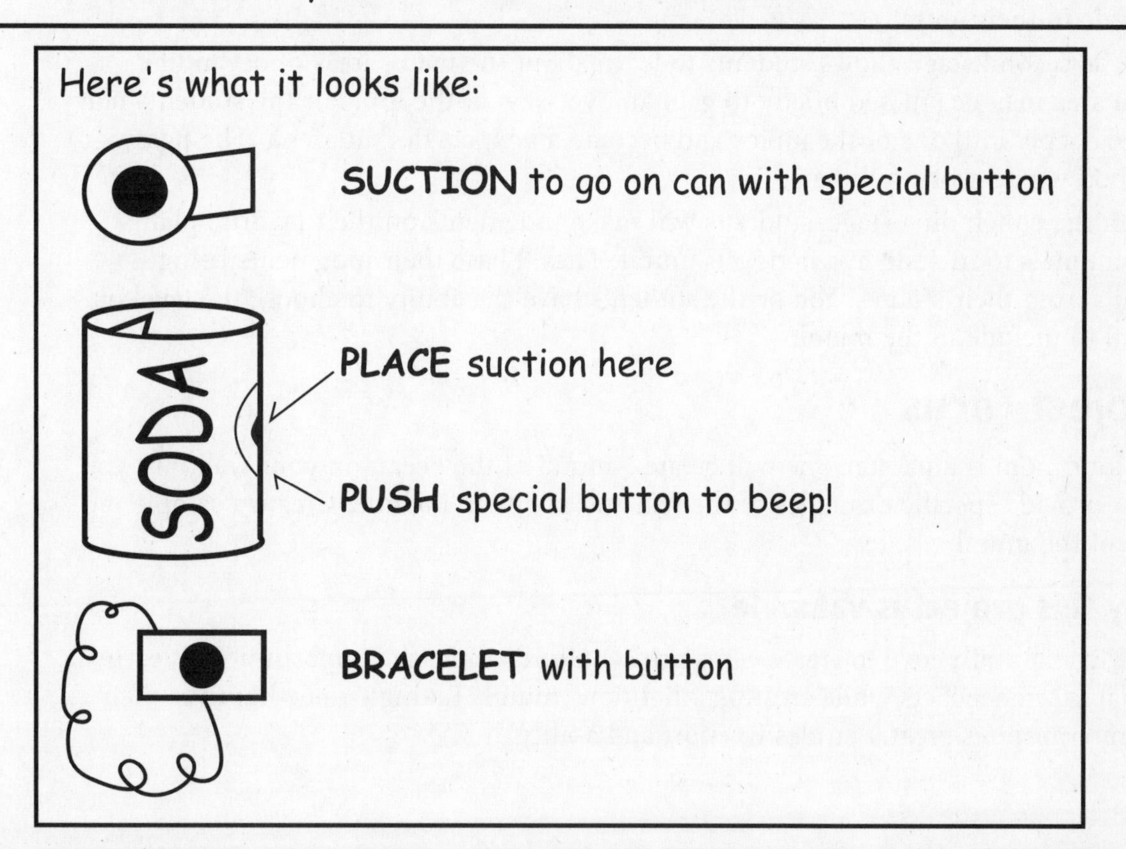

A sample proposal.

Chapter Six

Science
Grades 2-3

Ocean's Away

Oceans are huge mysteries filled with endless information for most students. By investigating specific elements students are able to break the vast resource of information into manageable pieces. Then by personalizing their studies to include interesting choices of their own, students relate oceans to their lives.

The first stage involves students building a basic foundation of ocean knowledge. Each activity is meant to build awareness of the many aspects of ocean life. Once students have a basic working knowledge, they will begin to form opinions about what they like and don't like. These opinions will help them when determining what to include in their mural.

The second stage allows students to learn about the many areas of ocean life. Each area may be studied briefly to gain an overview of the topic or the students may delve deeply into one of the topics and become an expert in that arena. The jigsaw method would work well here.

In the concluding stage, students will make judgments on their favorite plants and animals to include in a fictitious mural. They'll base their judgments using details from their studies. You or the students have the ability to choose the level of detail to include in the mural.

Project Focus

Your mom is a painter. She will create a mural of the ocean on your wall, but you must provide specific examples of the fish and plants to include. Create a sample picture of the mural.

Why this project is valuable

Kids naturally love to create. This project affords them the opportunity to learn valuable science facts while creating a fictitious mural. Taking ownership over their learning inspires greater strides in effort and ability.

Students will learn:

- Responsibility for their actions
- How to solve a problem
- Developing a useable plan
- Characteristics and facts of oceans
- Environment and habitats within oceans
- Ocean related jobs
- Ocean traveling vessels
- Mechanisms by which fish breathe

Vocabulary

- Tide
- Pacific Ocean
- Atlantic Ocean
- Artic Ocean
- Indian Ocean
- Currents
- Ocean floor
- Continental Shelf
- Gills
- Any specific fish or aquatic plant name

Teacher's Tip

There are many, many resources available for teachers regarding lessons about oceans. Find any that will help to move students in the direction you chose. Make sure to keep to the focus.

What kids should know before beginning this project

Students should have some knowledge of:

- Locating books in the library.
- Performing basic research in books and in the library.

Beginning the Project

Brainstorm and discuss

- Have students been exposed to oceans?
- Do students have basic knowledge of oceans?
- What are the four major oceans?
- What do the students know about oceans?
- Draw a web and map depicting the oceans of the world.

Chapter Six

Questions to consider

- What resources are available?
- Do students know any myths or legends about oceans?
- Do students have any ocean artifacts to share with the class?
- Could the art teacher have students make origami boats?

Activities

- Create project journal.
- Find five interesting facts about oceans.
- Write trivia questions for your facts.
- Create a display of various vessels that travel on oceans.
- Find five different jobs related to the ocean.
- Explain which of these jobs you would do if you had to in your journal.
- Make a list of different sports played on or around the ocean.
- Explain in your journal how to play one of these sports in your own words.
- Use a pumpkin, old ball, anything round to draw and label major oceans.

Developing the Project

Questions to consider

- Are there any resources specific to oceans available to you?
- Do the choices of students make sense?
- Have students detailed why they chose what they chose?
- Are there specific species needing to be studied?

Activities

- Make a list of five plants and ten species of fish to include in the mural.
- Name each choice and draw a picture of each.
- Develop the concept of how big the ocean is by comparing it to something.
- Write a joke, funny saying, or riddle about the ocean.
- Explain one way an ocean habitat is endangered.
- Draw a typical food chain in the ocean.
- Draw and label a plant and organism unique to each ocean.

Concluding the Project

Questions to consider

- Will any jobs be a part of the mural?
- Will any ocean vessels be a part of the mural?
- How many plants and animals will be included?
- Are the choices realistic?
- Do any students have fish related food allergies?

Activities

- Create the mural on paper.
- Make a 3D display depicting the mural choices.
- Act as an employer, explain an ocean related job to an applicant.
- Write a proposal to save the ocean habitat you discovered is endangered.
- Write a journal entry from the perspective of a fish—detail "A Day in the Life…"
- Hold an ocean life feast where students bring or cook various fish dishes.

Resources

Books

The Magic School Bus: On the Ocean Floor by Joanna Cole; Scholastic, Inc. (1992)
A Day Underwater by Deborah Kovacs; Scholastic, Inc. (1987)
At Home in the Tide Pool by Alexandra Wright; Charlesbridge (1992)

Internet Sites

http://www.seaweb.org—This valuable site has articles, facts and snippets of information on every aspect of oceans.

http://www.ocean98.org/—Another great site that offers great information on oceans.

http://www.yoto98.noaa.gov—This government site offers tons of fun facts and activities for kids to get involved in the information. Follow the link to the kids and teachers page.

Ocean's Away

Mural of the Ocean

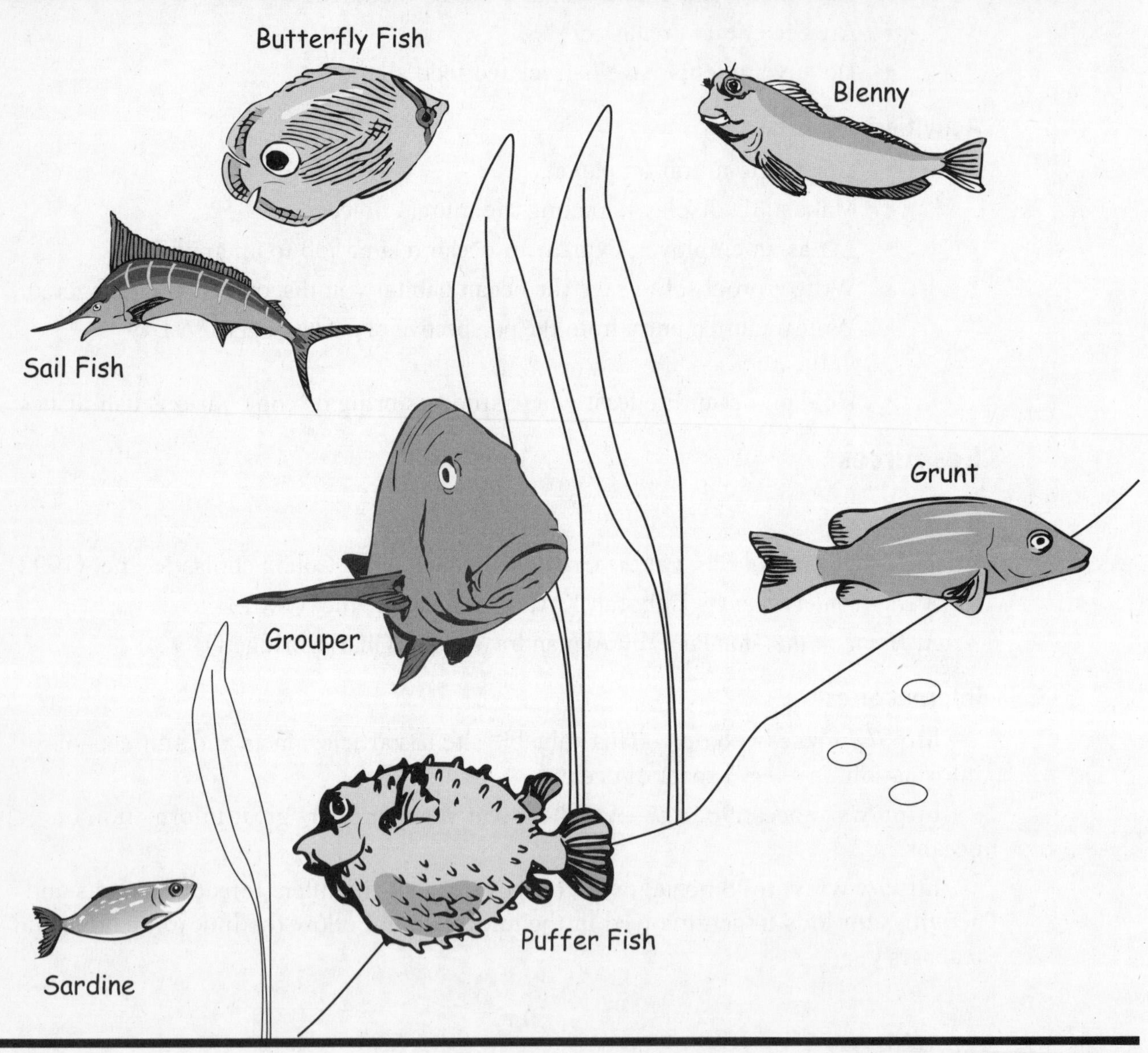

Sample of mural with fish labeled.

Science
Grades 3-5

Pesky Garden Creatures

Students enjoy learning about their environment. It helps them relate to their world. Learning key elements that aid in the growth of plants helps students develop an understanding of the science behind the symbiotic and parasitic relationships in a garden. It teaches students to understand balance in nature.

In the beginning stage of the project, each child develops an understanding of their local vegetation and insects common to the area. They will study how the climate affects local gardening and common names for local vegetation. They'll further learn the classification structure used to name all plants and animals. This structure is later used for all organisms included in the project.

During the developing stage, the class delves more deeply into specifics about the vegetation to their area. They will also investigate various relationships among organisms such as parasitic and symbiotic. Students study the scientists that research these organisms and the relationships within the organisms.

In the concluding stage, students must integrate the information to create a full garden design with specific elements included. These elements may vary depending upon curriculum goals and needs for your class. The garden design may include whatever creative elements you approve or deem necessary. Students will also learn about garden design and aesthetic factors in a garden.

Project Focus

Unwanted creatures are taking over your mom's garden. How can you remove the unwanted bugs without using chemicals?

Why this project is valuable

Students will ultimately be responsible for their environment. Learning to recognize imbalances in the environment and how to rectify these situations helps students build responsibility for their actions while learning about the delicate balances of plant and animal life.

Chapter Six

Students will learn

- Responsibility for their actions
- How to solve a problem
- Developing a useable plan
- Dynamics of an ecosystem
- Harms to an ecosystem
- Symbiotic and parasitic relationships
- Regional climate patterns
- Regional vegetation
- Regional gardening

Vocabulary

- Ecosystems
- Food Web
- Food Chain
- Parasites
- Symbiosis
- Plants
- Vegetation
- Entomology – Entomologist
- Botany—Botanist
- Binomial nomenclature

What kids should know before beginning this project

Students should have some knowledge of

- Vegetation and insects native to your region.
- System used in scientific names—Kingdom, Phylum, Class, Order, Family, Genus, Species
- Here's a great saying to help students remember the classification order: King Philip Came Over From Germany Saturday!

Teacher's Tip

As an extension for students that have a good working knowledge of ecosystems, have them study a fungus or disease that will overtake the garden rather than insects.

Beginning the Project

Brainstorm and discuss

- What is an ecosystem?
- What is a food chain?
- What is a food web?
- What is the climate of your region?
- How are organisms named?—Binomial nomenclature.
- Why do we use this system?—Create continuity across cultures.

Questions to consider

- How can students learn the precipitation and average temperature in your area?
- How do these statistics affect gardening in your region?
- What resources are available for research?
- Are there any experts on vegetation in your area?
- Could you take a field trip to study vegetation in your area?
- What is binomial nomenclature?

Activities

- Create project journal.
- Make a poster displaying climate for the region and affects on vegetation.
- Create a chart listing and showing common vegetation in the area.
- Create a chart listing and showing common insects in your area.
- Using paper clips connect pictures in a food chain authentic to your region.
- Use a specific organism and classify it scientifically through genus and species.
- Compare the scientific names of a brown bear and polar bear; a cat and a dog.
- Create a display depicting the various harmful elements to ecosystems—pollution, logging, urban development, oil spills, chemical spills, overpopulation etc.

Chapter Six

Developing the Project

Questions to consider

- What is the local vegetation?
- What are common bugs local to your area?
- Do any students garden regularly?

Activities

- Take a tour of a local nursery to learn about specific plants in your region. Write questions ahead of time for students to ask and know.
- Create a food web for organisms in your regions depicting the transfer of energy.
- Compare two specific smaller ecosystems—designate similarities and differences—i.e. a pond and a garden.
- Compare two larger ecosystems—designate similarities and differences—i.e.- tundra and prairie.
- Explain the difference between a symbiotic and parasitic relationship. Give examples of each.
- Explain what a botanist or entomologist does. Provide background and education needed to fulfill the job requirements.

Concluding the Project

Questions to consider

- Have students used realistic plants and animals?
- Were students able to obtain pictures and descriptions of each?
- Do students understand how to draw dimensions?
- Are dimensions realistic?

Activities

- Design a fictitious garden including complete dimensions and style, color and label diagram.
- Create a chart showing all vegetation and insects in garden, include scientific names and pictures.
- Create a poster, guide book, pamphlet, or multimedia presentation for your mom explaining how to rid her garden of pests.
- Compare the activities of a botanist and entomologist. What do they do the same/different? Explain which job you would prefer and why?
- Explain the steps in designing a garden.

Resources

Books

The Lorax by Dr. Seuss; Random House (September 1971) Although this does not deal with insects, this book provides a fun yet realistic upset to an ecosystem. Who can resist a Dr. Seuss book with a message?

Insects and Gardens: In Pursuit of a Garden Ecology by Eric Grissell; Timber Pr (November 2001) The author of this book is a research entomologist for the U.S. Department of Agriculture. He does a great job of explaining the ecology of a garden.

A Guide to Happy Family Gardening: A Little Help to Get Started Gardening with Kids by Tammerie Spires; Good Books (May 1999)

Internet Sites

http://www.gardenguides.com/—This site will provide facts and information regarding possible plants and insects found in typical gardens. Many articles are also available for perusal to build background information.

http://www.citygardening.net/index/—This site has many great tips and articles for city gardening, hence the title. Follow the link toward to middle of the page for common and scientific names of plants.

http://www.gardenersnet.com/atoz/insect.htm—This fantastic site provides pictures and information about common insects found in gardens. It also provides links to other valuable resources.

Pesky Garden Creatures

Renee's Garden

PLANTS		INSECTS	
Scientific Name	Common Name	Scientific Name	Common Name
Chysanthemum Maximum	Daisy	Certona Trifurcata	Bean Leaf Beetle
Saintpaulia Campanulata	African Violet	Plutella Xylostella	Diamondback Moth
Hedera Helix	English Ivy	Phyllotreta Striolata	Striped Flea Beetle

A sample chart for garden species.

Chapter Six

Science
Grades 3-5

Protecting Your Own

Students usually understand basic survival needs. Mostly they understand it in relation to themselves. By studying endangered animals and the factors that lead to their endangerment, students learn how to positively affect their environment to aid animals in their survival. Students can use their own survival needs to understand the needs of the animals.

In the beginning stage, students will learn the basic definition of endangered animals and many examples of these animals. They may choose to refine the study to animals in their state, region or country. Then students need to learn specific reasons and causes for endangering animals.

During the developing stage, students must delve deeper into the specific habitat of an endangered animal. They will develop an understanding of the habitat needs and resources available. Each child will recognize the relationship among the factors within the habitat that aid in the survival of the animal. Also, students need to recognize the specific factors that upset the habitat.

In the concluding stage, students should use the learned information to propose possible avenues and ideas for aiding in the survival of their endangered species. They may use a variety of resources and presentation methods for their proposal. Inclusion of scientific names and other pertinent habitat information may be included depending upon you curriculum goals and needs.

Project Focus

Every year certain animals become endangered. Pick an endangered species native to your state. Propose a plan to contribute to the growth of the species.

Why this project is valuable

Learning to respect and preserve the environment teaches students accountability for their world. Recognizing dangers close to home makes the threat more real for students.

Students will learn

- Responsibility for their actions
- How to solve a problem
- Developing a useable plan
- Habitats
- Habitat destruction
- Environmental protection
- Endangered species
- Food chains
- Food webs
- Chain reactions

Teacher's Tip

Studying multiple elements can become overwhelming for some students. For students that need help focusing, assign only one or two elements to study. Help them to specify the best areas to study and fine-tune their search.

Vocabulary

- Endangered
- Habitat
- Destruction
- Food chain
- Environment

What kids should know before beginning this project

Students should have some knowledge of

- Food chains
- Habitats

Beginning the Project

Brainstorm and discuss

- Read aloud *Who Eats What? Food Chains and Food Webs* by Patricia Lauber.
- Discuss structure of food chains.
- Draw sample food chains.
- Discuss habitats as homes with unique environments.

Questions to consider

- Are there any organizations in your state dedicated to endangered species?
- May students study animals outside of state, region, country etc.?
- If the subject is too broad, should students study just one element?

Chapter Six

Activities

- Create project journal.
- Write as many different definitions of endangered species as you can find. Then choose the one you like best and explain why in your journal.
- Make a list of five endangered animals you would like to learn more about.
- Conduct research to find which animal lives in or near your state, region, country etc.
- Read aloud *Will We Miss Them? Endangered Species* by Alexandra Wright.
- After reading aloud, make a list of the reasons why students will miss the animals, record in journal.
- Write a poem about your favorite animal.
- Make a list of common ailments to the habitat of five animals.

Developing the Project

Questions to consider

- How many factors are contributing to the endangered species?
- Do students need to focus on one aspect or several?
- Do students understand food chains and food webs?

Activities

- Draw your animal's food chain.
- Explain the habitat of each animal in the food chain.
- Draw a food web, including food chain of your animal.
- Make a list of common elements leading to the endangerment of your animal.
- Draw a picture of your animal's habitat.
- Read aloud *All the King's Animals: The Return of Endangered Wildlife to Swaziland* by Cristina Kessler.
- Make a list of steps taken to save animals in the book.
- Write in journal explaining if any of these steps could be used to save your animal.
- Create a list of possible ways to help your animal.

Concluding the Project

Questions to consider

- Were students able to obtain adequate information regarding animals?
- Any specific concerns regarding use of information?
- Has study spurred reaction in students?
- How do students feel about what they have learned?

Activities

- Propose a method or idea to contribute to the growth of your species. Write a report, present a slideshow presentation, or make a poster depicting idea(s).
- Write a letter to a friend describing the downfall of your species.
- Design an advertorial poster depicting ways people can help your species.
- Design a tri-fold brochure all about your species. Include habitat, food sources, eating and sleeping patterns, and other relevant information.
- Explain in your journal your feelings about what you have learned about endangered animals.
- Create a flow-chart of events that occurred to endanger your species.

Resources

Books

Who Eats What? Food Chains and Food Webs by Patricia Lauber; HarperCollins (1995)

Will We Miss Them? Endangered Species by Alexandra Wright; Charlesbridge (1992)

All the King's Animals: The Return of Endangered Wildlife to Swaziland by Cristina Kessler; Boyds Mills Press (1995)

Internet Sites

http://www.yahooligans.com/science_and_nature/living_things/animals/endangered_species/—This site could very well be the only site you need. There is tons of information, pictures, links and more.

http://investigate.conservation.org/—This acclaimed site offers specifics on conducting science fair projects and understanding biodiversity and conservation.

http://www.epa.gov/—The Environmental Protection Agency's official site. You can follow links to obtain information specific to your state and pages just for kids. There are also links to the Fish and Wildlife sites, including specific pages for kids and educators. Spend some time on this site.

Protecting Your Own

Kevin's Endangered Species

Protected as of March 11, 1967.
Habitat: 48 States, Canada, and Alaska.

Mid to late 1800s	Decline of eagles began with increased human contact.
1940	Bald Eagle Protection Act— Prohibited killing or selling of eagles.
After WWII	DDT and other pesticides decrease bald eagle population.
1967	Eagles officially listed as endangered.
1999	President Clinton announces "the Eagle is back."

A sample explanation of a species.

Science
Grades 4-6

Balanced Soil

Soil is often not thought of as anything significant. Surprisingly, a vast ecosystem exists just below our feet. Students are usually amazed to discover the delicate nature of this particular ecosystem. By using real examples of how soil affects the growth of plants, they are able to use their learning for authentic purposes. As with any environmental study, when students learn about their world, they also learn to place themselves within it.

For the first stage, students gain a basic understanding about soil. By examining soil and common properties of soil, they gain working knowledge of vocabulary terms and details specific to soil. Each fact aids in the overall understanding. Students also investigate the topography of their area. These details will be used when determining deficiencies and needs in soil.

In the developing stage, students personalize their learning by investigating specific properties of soil where they live. Each student furthers their understanding with soil tests and examinations. They will begin to examine composting as a natural fertilizer to address certain soil needs and will form judgments based on their findings.

During the concluding stage, students must make determinations of the soil in the area based on their findings. Then they use their findings to determine if specific plants or flowers will produce in that type of soil. Students may specifically look for plants that will grow in that soil or they may propose ways to enhance the soil to accept other flowers and plants. This would depend upon the needs of your students and curriculum.

Project Focus

You've just moved to the area. Your mom wants to plant her prized petunias. Is the soil right for petunias? If not, what should be done to make it right? (Could be any flower or plant)

Why this project is valuable

Understanding soil is part of understanding our world. Learning to identify soil properties enables students to develop planting techniques.

Chapter Six

Students will learn

- Responsibility for their actions
- How to solve a problem
- Developing a useable plan
- Effects of various elements on soil
- Layers of soil
- Nutrients and minerals in soil
- Transfer of nutrients in soil
- Composting methods
- Composting bins and supplies

Vocabulary

- Pedology—Soil Science
- Horticulture
- Nutrients
- Various flower or plant names
- Minerals
- Fertilizer
- Composting
- Topography
- Horizons

Teacher's Tip

This is a very broad subject. If time does not permit for each group to study each aspect in detail, break into parts such as: soil horizons, composting, and soil erosion. Use the jigsaw method to teach other groups about specialized areas.

What kids should know before beginning this project

Students should have some knowledge of

- All soil is not created equal.
- Various soil types.
- Topography of region.

Beginning the Project

Brainstorm and discuss

- Different types of soil.
- Different looks of soil. For example, the Red River along the Texas/Oklahoma border is very red-orange from the high clay concentration.
- Various uses of soil.
- What changes soil?
- What is a horticulturist?

Questions to consider

- Does your region have any specific soil concerns or limitations?
- Does your region have any specific vegetation issues?
- Does your region have any specific topographic considerations?
- Do students need a lesson in reading topographic maps?

Activities

- Create project journal.
- Create a display showing different soil colors and what each one means.
- Draw a topographic map of your region.
- Explain in your journal how the topography in your region would affect soil properties.
- Using garden tools, study a small area of soil. Dig to a depth of 6-8 inches.
- Examine with magnifying glass, draw and note: color, consistency, moisture level, plant and animal life, etc.
- Identify common minerals found in soil using internet, encyclopedia, or other resources.
- Explain composting in your journal. What does it mean? What does it do?

Developing the Project

Questions to consider

- How does erosion of soil affect the pH and mineral content of a soil?
- What do horizons look like?
- How thick is each horizon typically?

Activities

- Draw a picture showing the top three layers of soil composition. Explain what each level is. Tell one interesting fact about that layer.
- Conduct soil tests using a purchased kit from a garden center, mail order, or contact your county horticulture agent.
- Demonstrate specific details and findings of soil in a graph or table.
- Take a field trip to look at a cross section of land, along a cut roadway or any other property that shows the layers of soil.
- Compare different composting techniques. Suggest which is the best in your journal.
- Compare commercial compost bins. Suggest which is the best in your journal.
- Design a compost bin for your school. Include diagram.

Chapter Six

Concluding the Project

Questions to consider

- Do students have a good sense of soil in the region?
- Do students have a good sense of plants that grow in the soil?
- What possible resources could students use in developing end products?

Activities

- Provide a list of plants or flowers that can be grown in your soil.
- Create a display showing all growing preferences—light, water, pH, etc. of your mom's favorite plant.
- Draw on poster board a cross section example of your mom's soil explaining each horizon.
- Write a job description for a horticulturist explaining each aspect of his job.
- Create a display of the main erosion factors of soil detailing causes, include pictures.

Resources

- Contact a local farmer, landscape or agricultural expert. Many counties have a horticulture department that have information specific to your area. Check the local phone book.

Books

Gardening Wizardry for Kids by L. Patricia Kite; Scholastic Inc.(February 1995) This book has lots of great easy activities for kids to complete.

Start With the Soil: The Organic Gardener's Guide to Improving Soil for Higher Yields, More Beautiful Flowers, and a Healthy, Easy-Care Garden by Grace Gershuny; Rodale Press (May 1997)

Secrets to Great Soil: A Grower's Guide to Composting, Mulching, and Creating Healthy, Fertile Soil for Your Garden Lawn by Elizabeth P. Stell; Storey Books (February 1998) Field Guides on plants and flowers or any other gardening book.

Internet Sites

http://school.discovery.com/schooladventures/soil/—Discovery School site with kid friendly background on soil. This fantastic site has many fun activities like a *Soil Safari* and *Down and Dirty* for kids to adventure. It also offers teachers tips.

http://www.gardeners.com/—This site has several articles on soil development and pest and disease control. The site is written for gardeners but is easy reading for most students.

http://doityourself.com/garden/soiltexture.htm—This page contains information on an experiment to test soil texture with easy to read instructions for most students.

Balanced Soil

Mom's Prized Petunias

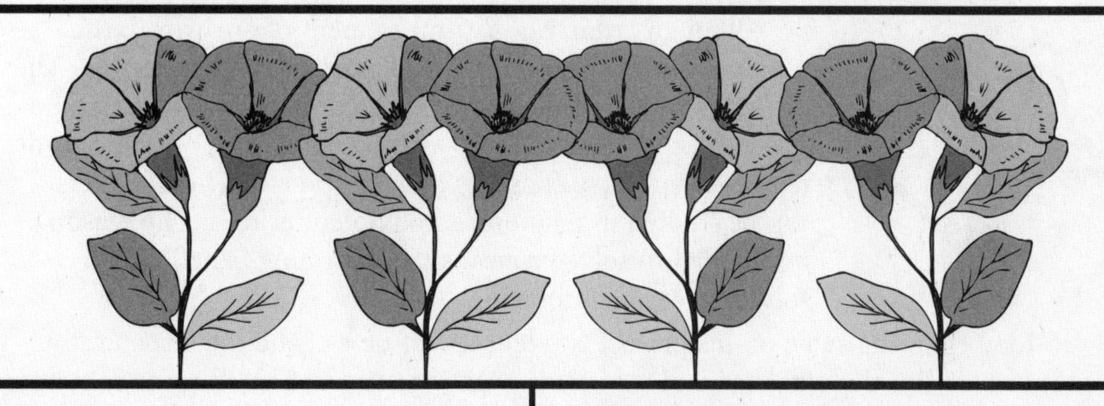

★ Sun: Full sun— outside or in.

★ Soil: Rich, loose soil.

★ Water: 1 to 2 times per week.

★ Type of Plant: Annual.

GROWING TIPS

—Add general purpose fertilizer to boost performance of bulbs and seeds.

—Water only when dry.

—Keep weeds out.

—Will not survive hard freeze.

A sample of statistics for a plant.

Chapter Six

Science
Grades 4-6

Crime Capers

Although crime is a natural element of our world, the intention of this project is not to glorify or exploit crime. The project is intended as an exercise in deductive reasoning and the science of crimes. Many interesting aspects of crime study require deep scientific skills and equipment. These aspects can be studied in relation to nonviolent crimes. The reasoning needed to solve crimes is the reasoning beneficial for solving everyday problems.

In the beginning stage of this project, students must investigate the steps used to report crimes and the persons involved in investigating crimes. They'll learn common traits of crimes and the sequence of events that lead to a crime being committed or solved. Each of these subjects will provide background for the students to further their investigation.

During the developing stage, each child will begin to personalize their studies by determining possible elements for their specific crime and crime scene. After deciding these factors, students must investigate the possible science concepts behind their crime and crime scene. Students should detail these with the evidence and clues for the case.

In the concluding stage, they will present a fully conceived crime and crime scene. The students may base the crime on facts learned through newspapers, television, or real crime data. Make sure students understand the nature of the crime is to remain nonviolent. Each novice should include evidence, clues, pictures and thoughts behind the crime. Some students may struggle with making all the pieces fit together while others are very detailed. The criteria may vary based on the needs of your students.

Project Focus

A non-violent crime has been committed. Explain and solve the case, including clues and evidence.

Why this project is valuable

Students will learn

- Responsibility for their actions
- How to solve a problem
- Developing a useable plan
- Deductive reasoning
- Process of elimination
- Criminal investigation field
- Forensic science
- Mysteries
- Story development

Vocabulary

- Crime
- Criminal
- Clues
- Evidence
- Forensics
- Suspect
- Perpetrator

What kids should know before beginning this project

Students should have some knowledge of

- What makes something a crime.
- The science behind crimes.

Beginning the Project

Brainstorm and discuss

- Breaking laws and crime.
- Non-violent crimes—provide examples to induce thinking.

Questions to consider

- What happens when a crime is committed?
- Are all crimes solved?
- Who solves crimes?
- What process do police follow when attempting to solve crimes?

Chapter Six

Activities

- Create project journal.
- Obtain several copies of the board game CLUE. Hold a contest on who can solve the case the fastest.
- Explain in your journal the process you used when playing the game CLUE.
- Write interview questions asking how to solve crimes.
- Interview someone in the criminal investigation field.
- Write a report or response sheet from the interview.
- Write a flow chart of steps to take to report a crime. Include what to tell police.
- Interview a forensic expert. Detail the science processes used in forensic science.
- Read a mystery novel or story.
- Watch an episode of Scooby-Doo. Detail what happened including the mystery involved, clues and evidence gathered and how the mystery was solved.

Developing the Project

Questions to consider

- What are the basics in investigating a crime?
- What key items should be searched?
- How would evidence be collected?
- How is evidence preserved?

Activities

- Design a crime scene check list. What questions would it include?
- Make a list of possible evidence pieces.
- Make a list of possible crime sites or areas.
- Make a list of possible non-violent crimes to be committed.
    - Create an outline of the crime: who, what, when, where, how and why.
- Draw a picture of the crime scene.
- Explain the science behind solving your crime. What are the clues?
- Research and report on a job related to the criminal investigative field.
- Outline the sequence of events from your mystery novel or story. Include clues and evidence characters used to solve crime.

Concluding the Project

Questions to consider

- Is the crime realistic?
- Is the evidence realistic?
- Any special materials needed?
- Is it a non-violent crime?
- Did students create logical conclusions?

Activities

- Provide a synopsis of the crime, evidence gathered, and how the crime was solved by the Captain of the Police Department in narrative or report form.
- Create an outline of a crime, including clues and evidence gathered.
- Set up a crime scene in the classroom. Work with partners to provide evidence and clues. Challenge other students to solve the crime.
- Write a report, present a poster, or use multimedia to demonstrate possible career opportunities in the criminal investigation field. Detail responsibilities and duties of each job.
- Design a different crime solving game. Include rules and game plan.
- Using plot theme from previous mystery story, write a mystery story of your own.

Resources

Books

Detective Science: 40 Crime-Solving, Case Breaking, Crook Catching Activities for Kids by Jim Wiese; John Wiley & Sons (February 1996) See list of books available on last website listed—provides extensive list of mystery books for students of all reading ability levels.

Internet Sites

http://www.nationalgeographic.com/ngkids/9902/crime-busters/index.html—Lots of great information and the opportunity to solve an animal caper.

http://www.fbi.gov/kids/6th12th/6th12th.htm—(For grades 6-12) The official FBI site catered to teach students about crime and job opportunities within the field of criminal investigation.

http://www.fbi.gov/kids/k5th/kidsk5th.htm—(For grades K-5) This site has interact text and provides a tour. Older students could still learn valuable information from this site.

http://www.bbns.org/ls/LIBRARY/booktalk/detectives/detectives.html—This site lists mystery books for early, middle and older readers.

Crime Capers

Roger's Crime Report

Crime	At 4:13pm, Roger walked outside to ride his bike. He saw his sister's teddy bear on the ground with tire marks on it, a torn arm and the red bow missing.
Suspicion	Someone ran over Roger's sister's teddy bear.
Clues and Evidence	Garage door is open. One car is missing. Dad said he needed to go to the store to get milk.
Suspects	Dad Grandpa—he has a key to Dad's car. Mom—she could have taken Dad's car.
Case Closed	Just as Molly, Roger's sister, was beginning to recover from her loss, Dad drove up with a red bow stuck to his tire.

A sample crime scene scenario.

Notes

Notes

Notes

Notes

Notes

Notes

Notes

Notes

Notes

A Parent's Guide to
First Aid

ISBN: 1-931199-20-5

Price: US$17.95 (CAN$26.95)

Nelson
A Parent's Guide to First Aid
parent's guide press

A Parent's Guide™ to
first aid
Roxanne Nelson, R.N.

- **When something goes wrong**, Parents and Caregivers are often the first ones on the scene. Knowledge of fundamental First Aid principles can be vital for the health of children. While nothing can replace qualified First Aid training, this book can provide an essential reference in time of need.
- **Common childhood mishaps** are listed in alphabetical order – from Allergic Reactions to Vomiting – for quick reference. Symptoms and causes are clearly explained by former pediatric and newborn intensive care nurse Roxanne Nelson to help caregivers understand the situation.
- **Step by step instructions** help caregivers determine the appropriate level of care and when to seek it – from a call to 911 in the direst circumstances or giving First Aid that can be undertaken at home.
- **Accident prevention** is the best First Aid one can give. The book includes common sense prevention tips for each situation, as well as general safety precautions parents can take around the home.
- **Chapters devoted to CPR/Rescue Breathing**, the Heimlich maneuver, and the emergency medical system each help parents and caregivers learn the basics before a crisis arises – helping them stay calm and in control of the situation.
- **What to stock in an effective home First Aid Kit**, where to keep it, and how to keep it up-to-date.
- **Indexed**.

parent's guide press

PO Box 461730
Los Angeles CA 90046
phone: 800-549-6646
fax: 323-782-1775